Remarkable Reptiles

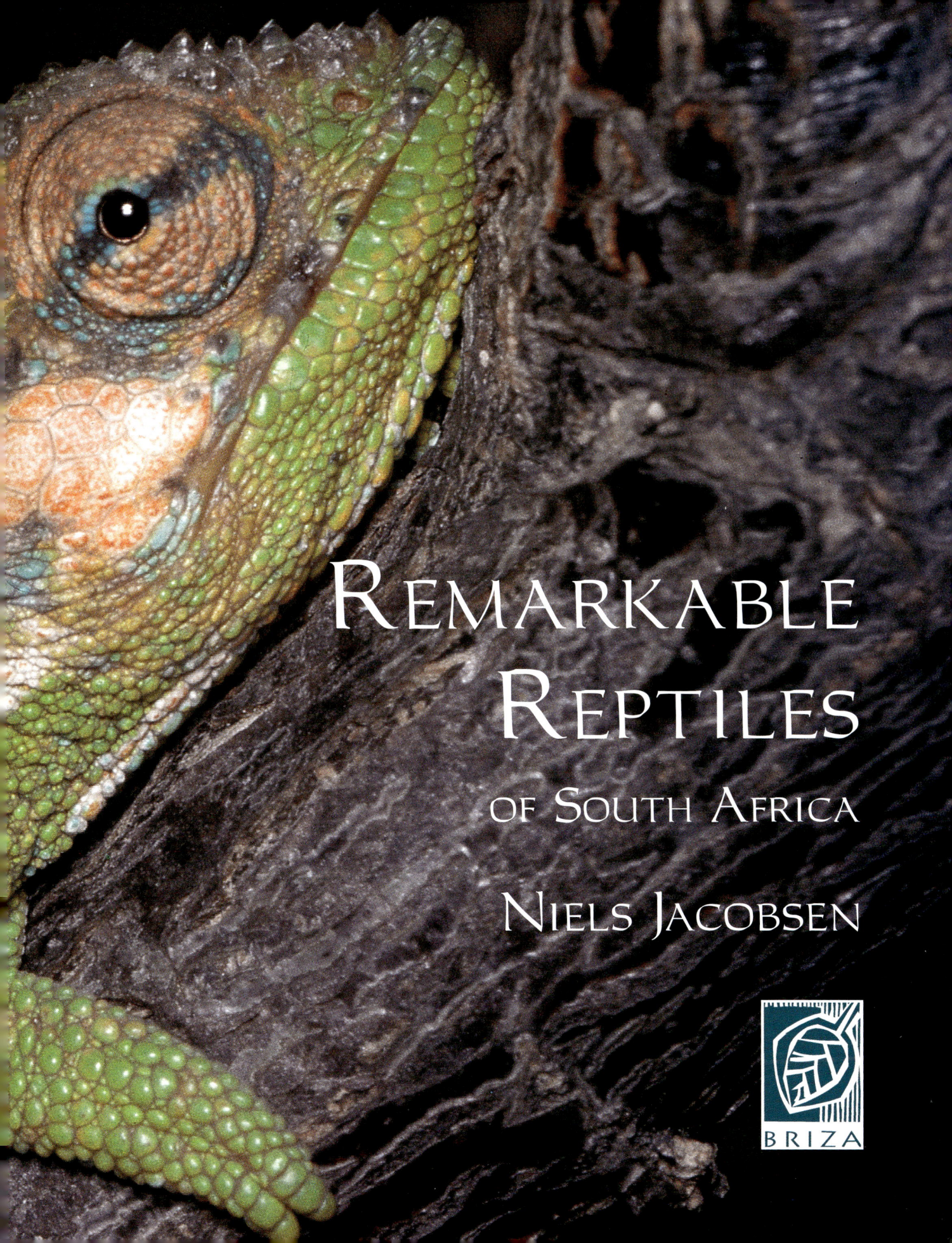

Published by
BRIZA PUBLICATIONS
CK 90/11690/23

PO Box 56569
Arcadia 0007
Pretoria
South Africa

First edition, first impression 2005

Copyright © in text: Niels Jacobsen
Copyright © in photographs: Niels Jacobsen and individual photographers mentioned
Copyright © in published edition: Briza Publications

All rights reserved. No part of this publication may be reproduced or transmitted in any form or by any means without written permission of the copyright holders.

ISBN 1 875093 49 4

Managing editor: Reneé Ferreira
Cover design: Sally Whines, The Departure Lounge
Design and typesetting: Alicia Artnzen, The Purple Turtle Publishing Services
Reproduction: Unifoto, Cape Town
Printed and bound by Tien Wah Press (Pte.) Ltd, Singapore

FRONT COVER: A male Transvaal dwarf chameleon showing off his bright colours.

HALF TITLE PAGE: A male Sekhukune flat lizard, endemic to Limpopo Province along the Leolo Mountains and ridges in Sekhukuniland.

TITLE PAGES: The Cape dwarf chameleon is one of the many endemic and attractive species in South Africa. It is restricted to the fynbos of the south-western Cape, with a translocated population occurring at Walvis Bay in Namibia. Photo: Richard Boycott

Contents

Introduction 1
Taxonomy and relationship to other animals 2
Reptile skin 5
Colouration 6
Skeleton 8
Hearing 9
Adaptations 10
Eyesight 12
Smell 13
Brain 13
Alimentary tract 13
Respiration and circulatory system 14
Thermoregulation 15
Hibernation 15
Reproduction 15
Growth and age 17
Zoogeography 17

Tortoises, Terrapins and Turtles 20
Order Chelonia
Skeleton 21
Food 22
Hibernation 22
Sexual dimorphism and reproduction 22
Enemies 22

Side- and Snake-necked Terrapins 24
Suborder Pleurodira
The Side-necked Terrapin family 24
Pelomedusidae

Hidden-neck Tortoises and Turtles 25
Suborder Cryptodira
The Land Tortoise Family 25
Testudinidae
The Pond Terrapin Family 32
Emydidae
The Soft-shelled Terrapin Family 32
Trionychidae

Sea turtles 33
Suborder Cryptodira
The Leatherback Sea Turtle Family 34
Dermochelyidae
The Modern Sea Turtle Family 36
Cheloniidae

Lizards 38
Order SQUAMATA
Suborder SAURIA or LACERTILIA
Thermoregulation 39
Anatomy 39
Smell 39
Caudal autotomy 40
Food 40
Behaviour 41
Reproduction 42
Longevity 43

The Agama Family 44
Agamidae

The Gecko Family 47
Gekkonidae

The Chameleon Family 54
Chameleonidae

The Skink Family 57
Scincidae

The Lacertid Family 62
Lacertidae

The Cordylid Family 66
Cordylidae

The Plated Lizard Family 69
Gerrhosauridae

The Monitor Lizard Family 71
Varanidae

Amphisbaenians 75
Suborder AMPHISBAENIA

The Worm Lizard Family 75
Amphisbaenidae

SNAKES 78
Order SQUAMATA
Suborder SERPENTES
   Adaptation 79
   Anatomy 80
   Thermoregulation 81
   Food 81
   Hearing 82
   Behaviour 83
   Reproduction 83
   Locomotion 84
   Diseases and enemies 85
   Myths 86
   Snake charmers 87

The Blind Snake Family 88
Typhlopidae
The Thread Snake Family 89
Leptotyphlopidae
The Python and Boa Family 89
Boidae
The Burrowing Snake Family 93
Atractaspididae
The Advanced Snake Family 95
Colubridae
The Cobra and Mamba Family 110
Elapidae
The Viper and Adder Family 121
Viperidae

CROCODILES 128
Order CROCODILIA
The Crocodilian Family 129
Crocodylidae

WHAT OF THE FUTURE? 138

BIBLIOGRAPHY 141

GLOSSARY 143

INDEX 146

ACKNOWLEDGEMENTS

The following persons are gratefully acknowledged for their friendly cooperation and assistance:
Dr J Kitching of the Bernard Price Paleontological Institute and the late Mr A.R. Hughes, Department of Anatomy, University of the Witwatersrand, for access to fossil reptile specimens.

A special word of thanks to the following people for generously making photos available, without which the book would be a lot poorer:
- Wulf Haacke
- Richard Boycott
- Johan Marais
- Lorna Stanton

Thanks are also due to all those people who have assisted in the compilation of this book, their efforts are appreciated.

Finally this book is dedicated to my wife Elsabie and all my colleagues with whom I have worked, and who share my concern for the survival of all that makes this planet heaven.

INTRODUCTION

This book is a general introduction to the reptiles in a South African context. To most people reptiles do not have the appeal of birds or mammals, and in fact many people are wary or even scared of them. The fact is that most reptiles are harmless, and in the case of those that are harmful the threat they present is usually blown out of proportion. Reptiles include tortoises, terrapins, turtles, lizards, snakes and crocodiles, their origins extending back into time long before the advent of birds and mammals. They are remarkable animals, which exhibit incredible anatomical and behavioural adaptations developed over many millions of years of evolution, enabling them to live almost anywhere on earth.

The ancestral lines of the living reptiles can be traced as far back as the Carboniferous period 315 million years ago, but owing to the paucity of fossil amphibian records, it is not possible to accurately determine what gave rise to the reptiles and when. Geological evidence points to wet environmental conditions at that time and one theory generally accepted today is that the original reptiles were aquatic animals which deposited their eggs on land. This was facilitated by the development of an amniotic egg, that is an egg in which the embryo is suspended in fluid retained within a membrane or amnion, in contrast to the amphibians which laid eggs in which only the yolk was enclosed by a membrane and which, like the eggs of fish, had to be deposited in water. Most animals at the time, including the predators, were aquatic or semi-aquatic and placing eggs on land would largely overcome predation. The era was characterised by floods and drought, and by avoiding the production of free-swimming larval stages

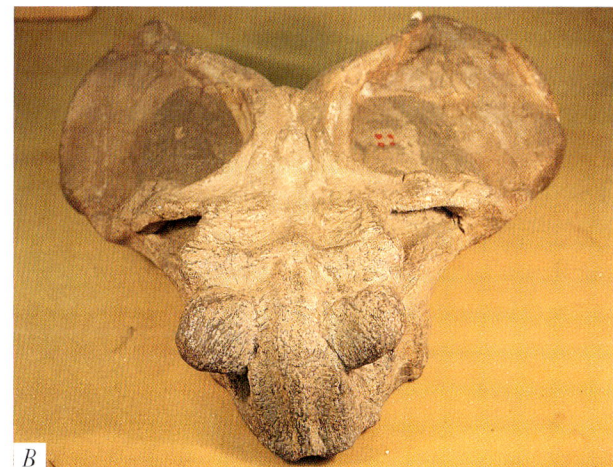

Reptile fossils from the past. Some of the primitive reptiles which roamed the Karoo about 190 million years ago: A. Aelurognatus, *a carnivorous mammal-like reptile;* B. Aulacephalodon, *a large herbivorous mammal-like reptile;* C. Thrinaxodon liorhinus, *a relatively advanced mammal-like reptile;* D. Prolacerta, *a primitive lizard.*

the reptiles reduced the risk of mass die-off due to dehydration or of being washed away. Many of the well-known early reptiles were semi-aquatic, which gives credence to this theory. However, who the actual ancestors of the reptiles were is the topic of several theories, some of which propose a polyphyletic origin involving two or more different groups of amphibians, which through parallel evolution developed the amniotic and cladoic eggs characteristic of the reptiles. However, as the taxonomical status and relationships of the various fossils is the subject of debate, no clarity can be given on this subject. It is generally thought that the Anthracosaurian branch of the amphibians, the Cotylosaurs, gave rise to the reptiles.

South Africa has a rich fossil fauna. Most people are acquainted with discoveries of the forerunners of *Homo sapiens* such as the hominids *Australopithecus africanus*, *A. robustus*, *Homo erectus* and associated mammals recorded from sites such as Sterkfontein and Makapansgat. What is less well known is that South Africa can boast of fossils of the first true representatives of the dinosaurs, tortoises, turtles and crocodilians such as *Mesosaurus* as well as the first mammal-like reptiles (Therapsids) such as *Aelurognatus*, *Aulacephalodon* and *Thrinaxodon liorhinus*, a branch of the reptiles which gave rise to the mammals. The fossil record of the latter is among the most complete in the world, examples of which are housed at the Bernard Price Institute of Palaeontology, University of the Witwatersrand. Both herbivorous and carnivorous species dominated the Karoo some 250 to 190 million years ago. Other branches of primitive reptiles also roamed South Africa, including prehistoric lizards, such as *Prolacerta*. Most of our knowledge comes from fossils found in shale beds underlying the Karoo and in the sandstones of the Eastern Cape and Free State where many were preserved in sedimentary rocks derived from the deposition of water or wind borne sand and silt, which covered animals trapped in the mud of drying lakes, pans and rivers or those killed by predators, becoming fossilised in time.

However, the living reptiles of today only represent four of the 17 Orders which roamed the earth during the Mesozoic era some 200 million years ago. These include the Chelonia, Squamata, Rhynchocephalia, Tuatara and Crocodilia, of which only the first two and the last are represented in South Africa.

Taxonomy and relationship to other animals

What is a reptile and how does it relate to the other groups of animals we know today? The whole animal kingdom is subdivided into groups or classes, each with similar structural and anatomical features. These can be clearly defined on the grounds of such features into:

Phyla (singular Phylum, a word derived from the Greek word *Phulon* meaning a tribe) each of which is divided into Classes, each of which is divided into Orders, which are then split up into Families. Families are subdivided into Genera and these are comprised of Species. This arrangement is usually subdivided for convenience to include Subphyla, Superorders, Suborders, Subfamilies and Subspecies and even Varieties. All these elaborate particular morphological, anatomical, genetic and behavioural characteristics of the various animals serve to separate them from others with similar but not all of the same characteristics. This enables us to classify different animals and arrange them in an orderly manner so that relationships can be easily recognised.

Reptiles belong to the Phylum Chordata as do all animals with a vertebral column or spine. This is the first major division and it separates these animals from the insects, molluscs, crustaceans and arachnids, which are termed invertebrates as they do not have such a structure. This vertebral column is composed of a number of jointed bones surrounding and protecting the spinal cord. Apart from this, reptiles also share with other vertebrates structures such as ears, eyes, nose and skull, a heart, liver and kidneys. They therefore function very much like we do with certain exceptions. Reptiles are commonly called cold-blooded, which actually has little to do with the blood but simply reflects with a few exceptions their inability to generate heat physiologically as is the case with birds and mammals. They are dependent on the sun and ambient temperature, adapting their behaviour to this in order to be able to function properly. In addition the temperature at which the eggs incubate, determine the sex of crocodilians, chelonians and some lizards. In crocodilians high temperatures result in males while in chelonians this produces females. Lizards vary depending on the group.

The Phylum Chordata is subdivided into Classes including Aves (birds), Mammalia (mammals), Pisces (fish), Amphibia (amphibians) and Reptilia (reptiles). On account of the gradual transition from amphibians to reptiles and from reptiles to birds and mammals many features are retained by each class, provided they are advantageous to the evolving group. Birds, for instance, are simply highly specialised reptiles. However, in spite of these basic anatomical likenesses there are many differences between reptiles and other vertebrates, including a skin of keratinised scales or plates, a single condyle at the base of the skull on which the head swivels and a lower jaw each half of which is composed of several bones.

Depending on whose viewpoint one follows, the Reptilia are divided into a varying number of subclasses. The basic layout (see box overleaf) follows that of Bellairs (1969) who followed a classification by A.S. Romer, amended according to the latest evidence. In more recent times many changes have been made according to more definitive techniques, some of which have been incorporated here. There are three subclasses of living reptiles, namely Anapsida, Lepidosauria and Archosauria, but only the first and last are present in South Africa.

The green-eyed Marbled tree snake is a nocturnal species occurring in the lowveld of Mpumalanga and Limpopo provinces, Swaziland and northern Zululand.

A classification of South African reptiles

Subclass Anapsida
Order Chelonia or Testudinata (tortoises, terrapins and turtles)
Suborder Pleurodira
Family Pelomedusidae (side- and snake-necked terrapins)

Suborder Cryptodira
Family Emydidae (typical terrapins)
Testudinidae (land tortoises, pond and box terrapins, etc.)
Cheloniidae (sea turtles)
Dermochelyidae (leatherback turtle)
Trionychidae (soft-shelled turtles)

Order Squamata
Suborder Lacertilia or Sauria (lizards)
Family Agamidae (agamas, etc.)
Chamaeleontidae (chameleons)
Gekkonidae (geckos)
Scincidae (skinks)
Lacertidae (sand lizards, mountain lizards, etc.)
Cordylidae (girdled lizards, snake lizards, flat lizards)
Gerrhosauridae (plated lizards)
Varanidae (monitor lizards)

Infraorder Amphisbaenia
Family Amphisbaenidae (worm lizards)

Suborder Ophidia or Serpentes
Family Typhlopidae (blind snakes)
Leptotyphlopidae (thread snakes)
Boidae (pythons)
Atractaspididae (African burrowing snakes) (Subfamilies Atractaspininae, Aparallactinae)
Colubridae (Subfamilies Lamprophinae, Natricinae, Colubrinae, Psammophinae, as well as the specialised Subfamily Dasypeltinae)
Elapidae (cobras, mambas) (Subfamilies Elapinae, Hydrophiinae (sea snakes))
Viperidae (adders, pit vipers, rattlesnakes, vipers) (Subfamilies Viperinae, Crotalinae)

Subclass Archosauria
Order Crocodilia
Suborder Eusuchia
Family Crocodylidae (crocodiles, gavials, alligators, caimans) (Subfamilies Crocodylinae and Alligatorinae)

Reptile skin

One of the most obvious characteristics of reptiles is the skin or integument, which was an all-important development for an animal emerging from an aquatic or semi-aquatic environment to take up life on land. Amphibians all have porous skins and die rapidly from dehydration, even forms which are to a large extent independent of surface moisture. The integument of reptiles does not have to be kept moist in order to facilitate respiration. All reptiles have epidermal scales, which are formed by the folding of the outer skin layer (epidermis) and the deposition of keratin, which forms a waterproofing layer. The latter is also present in mammalian skin as well as being a major constituent of hair, horns and feathers. This material is dead and constantly flakes off or is rubbed off. Continual replacement takes place from below by living epidermal tissue. In lizards and snakes removal takes place several times a year and may be in the form of large flakes or as a continuous slough, whereas in tortoises and crocodiles it flakes off very much like that in mammals. When snakes or lizards are ready to slough the body, including the eyes, is covered in a bluish-white opaque skin, which frequently causes a snake to be irritable at this stage as it cannot see and is therefore vulnerable to predation.

The mechanism of sloughing is a complicated and involved process initiated by the diffusion of lymph between the old and new skin layers, causing nerve and circulatory connections to be severed. This loosening of the skin around the body causes the opaqueness which obscures the natural colour of the underlying new skin. An enlargement of the head through an increase in blood pressure in the veins of the head, caused by muscular contraction surrounding the internal jugular vein, both loosens and ruptures the dead skin layers. It is then rubbed off against an object such as a rock or branch.

In most lizards removal is piecemeal and frequently the skin is consumed during or after removal, which is often aided by biting and scratching. In snakes the skin is mostly shed intact but inverted as it is rolled off from head to tail tip.

Scales vary in size and structure from small granular scales to large spiny structures such as those found in the lizard family Cordylidae (girdled lizards) and occur in definite patterns and arrangements which are very useful in identifying different species. Many lizards, such as plated lizards of the family Gerrhosauridae and crocodilians, have small plates of bone called osteoderms in the dermal part of the scales, which make the skin exceedingly tough, encasing the body in armour.

Apart from the waterproofing function, many reptiles display specialised modified scale structures. The elongated scales above the eyes of some of our small adders, such as the Horned adder, are possibly designed to enhance the fierceness of expression of the head. The large rostral shield of the Shield-nosed snake functions as a battering ram when burrowing. The burrowing worm lizards have similar structures as an aid to their fossorial existence. The heavy spiny scales of girdled lizards function to anchor these lizards in a burrow or crevice should attempts be made to remove them from their retreat. At the same time the armoured tail is thrashed about in defence.

Probably the best-known adaptation is that of the rattle of rattlesnakes, which consists of modified scales and is used to warn off potential enemies. The broad ventral scales of snakes are essential in locomotive ability.

Contrary to popular belief the skin of reptiles is not slimy but completely dry, although in some species the scales are so smooth as to give a very shiny appearance,

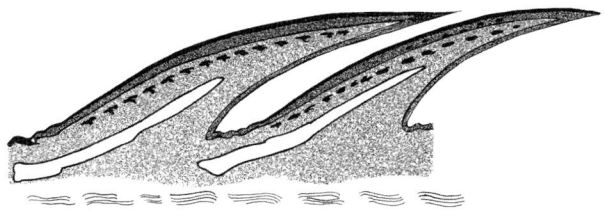

A longitudinal section through snake or lizard skin. Note the position of the melanophores or black pigment cells which underlie the other pigment cells.

Osteoderms form a protective armour around the body of this Giant plated lizard, Gerrhosaurus validus, *even after the death of the lizard. Only the soft tissue decomposes, leaving the bony shell.*

such as that of the Cape cobra. There are no mucous glands like that found in amphibians or any sweat or grease glands as found in mammals. There are a few glands in localised regions of the skin, such as scent glands in the cloaca, which may be used as a sexual attractant or as a deterrent to would be predators. Some lizards, including geckos and girdled lizards, have preanal and femoral pores in males. In other species they are found in both sexes but are usually enlarged in males. These are filled with glandular secretions but their function is unknown although probably of sexual significance. Some turtles have highly developed musk glands along the edge of the lower jaw and sides of the body. Most notable is probably the Cape terrapin, which has a very strong odour if disturbed or picked up. This noxious odour no doubt assists in defence.

COLOURATION

Most reptiles are cryptically coloured according to whatever environment they inhabit. Therefore, green snakes usually inhabit trees, shrubs or are found along the banks of streams where the vegetation is usually green. Sandy or brown snakes are found in arid or rocky environs. Their colours are therefore a function of their habitat and mode of life. Very few species display aposematic colouration as is frequently found in frogs and which serves as a warning to potential predators that they are poisonous. The American coral snake is one such species and its bite is poisonous, a fact which is mimicked by the Sonoran king snake, a harmless species, by adopting a similar colouration. The males of many reptile species, especially lizards, exhibit very bright colours which serve a dual function, namely to attract a female and as a display of dominance, acting to deter other males of the same species from possession of the female or of a territory.

Some species of lizards are capable of varying degrees of colour change depending on the colour of their surroundings or the mood of the individual. Amongst the most familiar are the chameleons, but other species include geckos and agamas although not to the same extent.

Although a brief discussion of the mechanism of colour change in chameleons is included later, it is perhaps necessary to go into some detail here. The integument of lizards shows a variety of pigment cells or chromatophores. Some species only have melanophores, containing a black pigment, others include melanophores together with guanophores, or both of these together with lipophores, or including allophores. The pigment cells occur in a range of combinations. The cell bodies of melanophores are found in the dermis beneath all the other chromatophores. They have branching processes which extend outwards towards the basal cells of the epidermis. As a rule the melanophores form a dense and almost continual sub-epidermal layer and are responsible for dark colouration. Lack of, or small quantities of melanin results in light coloured individuals, of which perhaps the most notable are albinos. Lipophores and allophores contain yellow and red pigments respectively. They lie immediately below the epidermis and above the melanophores. The production of yellow or red pigments produce these colours as well as various shades of brown when their effects are combined with those of the melanin producing cells. Blue and green pigments appear to be absent in reptiles. Although these colours are prevalent in many species of snakes and lizards they are formed by special cells called guanophores, which apparently contain a semi-crystalline material called guanine and is related to uric acid. The guanophores lie in the upper dermal layers. The guanine is virtually colourless and is a good reflector, modifying light that falls on it. The guanophores by themselves produce a white effect but in combination with melanin producing cells, scatter light of certain wavelengths to produce a blue colour. If there is a layer of yellow pigment cells between them and the light source the colour produced will be green. Pigment cells are not only restricted to the skin but may

Colour change in the Common flap-necked chameleon, Chamaeleo dilepis, *pale at night when the animal is sleeping and the threat from predators is low, and dark during the early morning while basking to increase its temperature and mobility.*

also be present in membranes lining the mouth, the body cavity and in the mesenteries of the viscera. For instance the inside of a Black mamba's mouth is blackish, that of the chameleon yellow. The dark viscera of some species may protect the animal from excessive heat and light radiation.

Some reptiles produce a colour change by expansion of the skin brought about by inflating the body or throat with air. This exposes the interstices between the scales, which may often be blue or some other bright colour. This is mostly used as an aggressive or threat display among lizards and snakes, including chameleons. However, the colour change most familiar to people is that of the chameleon and agama. In these creatures colour change is effected by the dispersal of brown to blackish melanin pigment in the melanophores.

Generally speaking there are two responses that induce colour change – one depends on the degree of lighting of the background, while the other depends on the mood of the animal. There are, however, several other factors involved including light intensity and temperature. Mechanisms of colour response are mainly initiated through the eyes, but the skin is also light sensitive, although not to the same extent. It is interesting to note that chameleons are paler in colour at night and show up in the light of a torch against the darker background of the foliage.

There are two possible explanations for the mechanism of colour change. One is that of nervous control whereby the retina of the eye is stimulated by light, which initiates a train of nerve impulses via the optic nerve to the brain, down the spinal column and through the peripheral nerves to the melanophores. The other is glandular, whereby the pituitary or adrenal glands are stimulated to produce hormones which are then circulated by the blood. Stimulation of the glands may be nervous, initiated by the retinal cells of the eyes. The production of adrenalin may be responsible for the induction of excitement pallor. Chameleons, however, appear to rely on nervous stimulation and this will be discussed in more detail in the relevant section.

SKELETON

Like most vertebrates, reptiles have an internal skeleton composed of bone and cartilage, which supports the body. The degree of cartilaginous material varies and is most prevalent in young animals. It is gradually replaced by bone with increasing age. In the adult, cartilage remains in parts of the skull and shoulder girdle (where these are present) and over the surface of the joints. Bone growth continues throughout life, although in many instances the rate of growth decreases with age. This is in contrast to mammals where growth ceases at the attainment of a certain age.

Reptiles are a very diverse group and occupy many different habitats as a result of a range of morphological adaptations. There is regional differentiation found along the vertebral column. Cervical vertebrae with short ribs which do not reach the sternum and which may even be absent altogether are found in all reptiles. As is the case in birds and mammals, the second vertebra forms the axis. There is little differentiation between thoracic and lumbar vertebrae and all have ribs attached, although they may be shorter posteriorly and not as flexible.

In snakes the ribs are attached to the large ventral scales and assist in movement. In chelonians, crocodilians, tuatara and most lizards, limbs are the typical mode of progression. Marine turtles have modified limbs adapted to swimming while many lizards and all the worm lizards show various reductions in limb development as a result of adaptations to specific habitats or niches.

The skull of reptiles varies according to the Order it belongs to. The differences are mainly in the number of temporal openings and where these are situated. Chelonians do not have a temporal opening (anapsid), which places them in the Order Chelonia.

Most living reptiles fall into the diapsid type, where two temporal openings are divided by a narrow bar of bone to a greater or lesser extent, but at least usually present in the embryo. There is, however, considerable variation in this pattern, which forms the basis of differences between Orders.

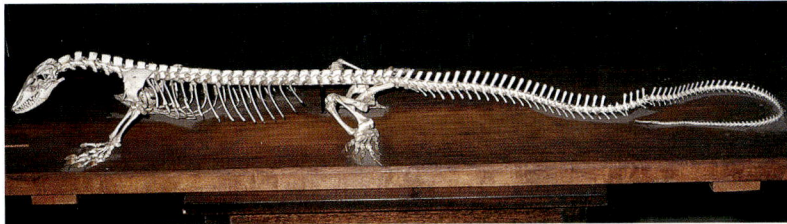
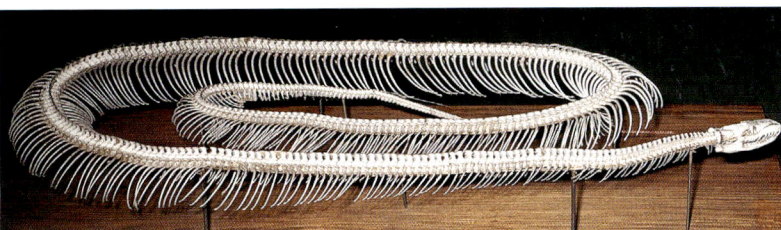
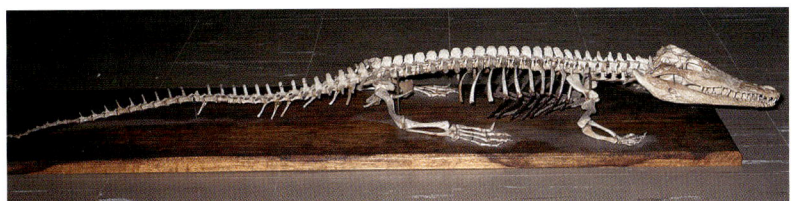
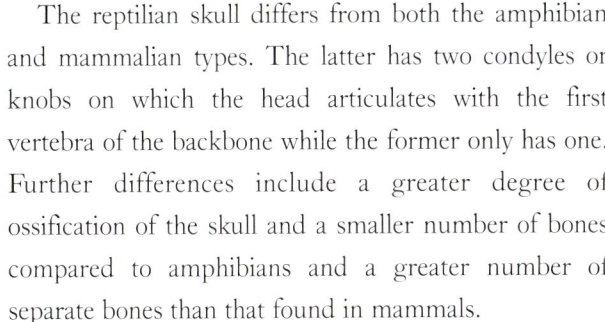
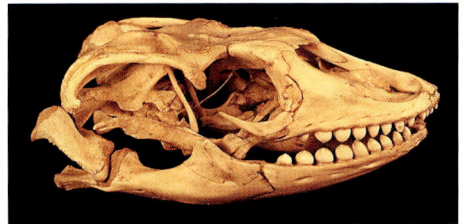

ABOVE:
The skull of the Veld monitor, Varanus albigularis. *Note the teeth which are adapted to grip and crush prey. The large number of bones is typical of most reptiles.*

LEFT, TOP TO BOTTOM:
Skeleton of a Water monitor, Varanus niloticus. *The long tail is used to propel the animal through the water.*

The skeleton of a Southern African python, Python natalensis. *Note the elongated ribs, which are attached to the ventral scales permitting rectilinear or caterpillar-like movement*

Crocodile skeleton.

The reptilian skull differs from both the amphibian and mammalian types. The latter has two condyles or knobs on which the head articulates with the first vertebra of the backbone while the former only has one. Further differences include a greater degree of ossification of the skull and a smaller number of bones compared to amphibians and a greater number of separate bones than that found in mammals.

An interesting feature of some reptile skulls is cranial kinesis, which refers to the movement of the upper jaw on the braincase. This is not found in crocodiles, turtles and the tuatara but is present in varying degrees in lizards and advanced snakes. This has the advantage that the animal can open its jaws wider to accommodate and manipulate larger prey. In snakes this has developed to a marked degree with the jaws and palatal structures loosely attached to the skull, allowing each side of the skull independent movement. The lower jaw is made up of several bones, similar to that found in birds and amphibians but not in mammals where the lower jaw is formed by a single bone on each side.

HEARING

A feature common to reptiles, birds and amphibians is the hearing mechanism, which consists of a single bone, the stirrup or stapes, whereas mammals have three – the stapes, malleus and incus. Little is known of hearing ability in reptiles but most lizards are able to hear airborne sounds of low frequency. Among the more vocal lizards are geckos, many of which utter a variety of clicking sounds. The best-known indigenous species are the so-called barking geckos, which occur in the arid parts of South Africa and call actively at dusk.

Most lizards have external eardrums, in contrast to snakes, which have none but are apparently able to hear sounds of very low frequency. Snakes do not communicate by sound – the hissing sounds they produce when disturbed are caused by the violent expulsion of air from the lung. Although there is little conclusive evidence it is generally believed that snakes are able to pick up vibrations from the ground by contact with the jawbone. The vibrations are said to be transmitted via the quadrate bone, which is in contact with the stapes and the jawbone.

Crocodiles and alligators are known to utter a variety of sounds and are the only reptiles to have developed an external flap-like outer ear. This flap seems to be mostly used to protect the eardrum from possible injury by underwater snags or even pressure when the animal dives. Normally there is only a slit left open at the anterior end when the head is out of the water but this closes when submerged.

Some chelonians, in particular terrestrial species, are known to be able to perceive low frequency sounds and males utter a variety of sounds during mating. Tortoises also hiss by expelling air though the nostrils. The ability to hear is however inferior in tortoises and especially in marine turtles when compared with that of lizards and crocodiles. Sounds therefore play little part in chelonian life except in the mating season when cries and grunts are sometimes heard.

ADAPTATIONS

Many of the adaptations of reptiles relate to environmental conditions. Snakes for instance exemplify this adaptability. Large-bodied snakes such as pythons, boa constrictors and Puff adders, because of their heavy bodies, move in what is termed rectilinear locomotion. This type of movement is effected by the ventral body scales moving forward by muscular activity and hooking onto projections on the ground, pulling the animal forward. Most snakes progress by means of lateral undulations of the body, whereby the bends of the body push and pull the snake along. Snakes cannot move fast over relatively smooth surfaces and are, for instance, almost helpless when crossing tarmac roads. A third method of locomotion is that of the side-winding practices of desert adders, which probably evolved in response to movement across a highly unstable sand surface.

In lizards some interesting and peculiar mechanisms have evolved. Many geckos have the tips of the toes expanded and covered by horizontally layered pads, called lamellae. These are covered by minute hairs, which enable the animal to cling to vertical glass panes or even hang upside down from the ceiling, although this ability is further assisted by the presence of a small claw near the tip of the digit.

The adaptation of the chameleon to an arboreal existence is just as remarkable. The digits are arranged for gripping rounded branches and the feet are flexed even when asleep. It is therefore all the more remarkable that a few species (only one in southern Africa) have turned away from the trees and have become terrestrial again as their ancestors probably were.

Other highly specialised adaptations include the feet of the desert-living Web-footed gecko, *Palmatogecko*

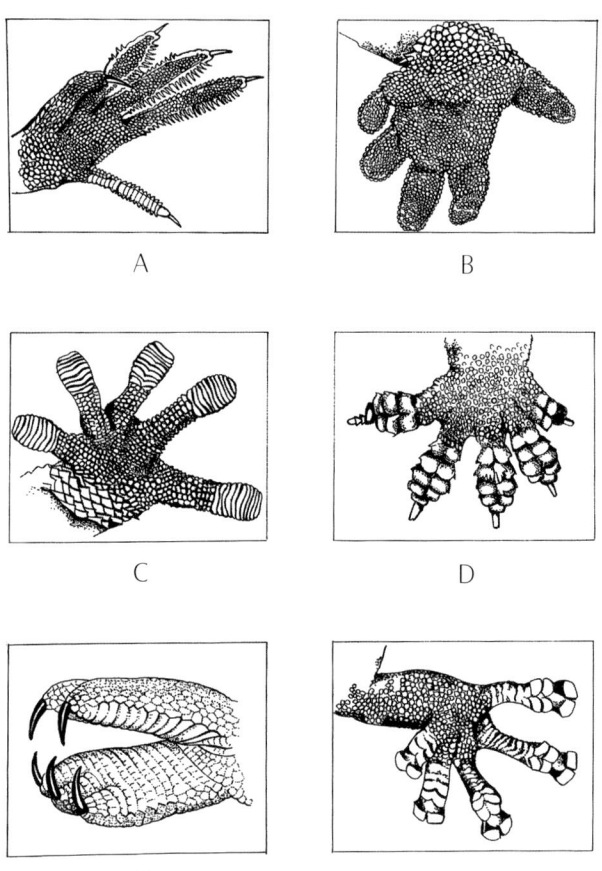

Adaptations of lizard feet: A. Barking gecko, the elongated scales helping the lizard to move swiftly across loose sand; B. Giant ground gecko, the large unspecialised feet are adapted to a variety of soil surfaces; C. Turner's thick-toed gecko, the lamellae under the toes help the animal to climb on rocks and large termitaria; D. The feet of the Tropical house gecko are adapted for climbing trees and the walls of houses; E. Common chameleon, the opposing toes enable the lizard to maintain an exceptional grip on branches; F. African flat gecko, the lamellae are adapted for running across rocky surfaces.

rangei, of the Namib, which are webbed and assist it to cross unstable sand surfaces and even to excavate burrows. Other sand-dwelling lizards have developed elongated scales or hairs to facilitate movement on loose surfaces. Another unusual development in overcoming restrictions in movement imposed by the habitat is that of some species of forest-living lizards and snakes found only in Asia. They have developed elongated ribs which can be expand and enable the animal to glide across open spaces from tree to tree thereby avoiding the necessity of having to climb up and down trees 20 to 30 m high. While no South African lizard is bipedal, this form of locomotion has been developed by some agamids and varanids from Australia and iguanids in Central America. These animals run on their hind feet using the tail as a balancing organ. This adaptation has resulted in the development of a narrower pelvis in some species. Familiar to most of us are the Water monitor and Nile crocodile, which are adapted to an aquatic existence and swim by using the tail with lateral undulations, propelling them rapidly through the water. The limbs are held adpressed to the sides of the body to streamline it as much as possible. One species of lizard (an iguana) and the sea snakes have returned to the sea. The latter have developed cylindrical bodies, in many cases with laterally flattened tails which are used as swimming organs. Thus an amazing diversity of forms has developed in association with the establishment of new niches and habitats.

Another skeletal development in reptiles concerns the presence of two sacral vertebrae as opposed to one in amphibians, followed by a distinct tail. The tail in most lizards and the Tuatara can be detached voluntarily along a fracture plane in a process called autotomy, which will be discussed in greater detail in the chapter on lizards. It does, however, result in the loss of the caudal vertebrae below the fracture. In amphisbaenians many species

The webbed feet of the Web-footed gecko are an exceptional adaptation for movement across unstable sand surface.
Photo: W.D. Haacke

display this fracture zone externally, demarcated by a slight constriction of the caudal annuli (rings of scales). It is an excellent defence measure particularly as the severed tail continues to wriggle about for some time afterwards drawing the attention of a predator away from the animal enabling it to escape.

While regeneration of the tail takes place in most lizard species, some do not have this ability and twist off the tail. This results in permanent damage with no regrowth. Such self-mutilation has also been recorded in snakes especially some sand snake species which have a habit of twisting off the tail if held near the tip. The severed tail continues to writhe about attracting the predator's attention while the snake escapes. It does not regrow again.

The pectoral and pelvic girdles must be mentioned here, as they are characteristic of most reptiles. Many snakes, however, have no pectoral girdle due to their complete reduction in limbs. A few families, such as the Typhlopidae, Leptotyphlopidae and Boidae, have retained a vestigial pelvic girdle, which is mostly internal, external evidence being limited to reduced digits or claws found in pythons. These have no function, and are remnants of their lizard-like ancestor.

EYESIGHT

Reptiles rely mainly on their eyesight and sense of smell to locate prey, perceive danger and to orientate themselves in their surroundings, with the exception of burrowing snakes such as blind and thread snakes and some burrowing lizards, which have degenerate eyes and possibly only perceive differences between light and dark. Many diurnal lizards and snakes cannot see at night while many nocturnal species cannot see in the light. Diurnal species tend to have rounded pupils and possess a fovea centralis as opposed to nocturnal species, which have slit-like, vertical pupils and the fovea centralis is absent. Many nocturnal geckos have slit-like pupils, often with a series of pinhole openings.

The best eyesight of all snakes is attributed to the vine snakes, which have horizontal keyhole- or dumbbell-shaped pupils and a pointed head with grooves along the side of the snout, allowing for exceptional bifocal vision even to the extent of detecting stationary prey. Of particular interest is the fact that both diurnal lizards and snakes have yellow filters in the eyes to screen out shorter wavelengths and to reduce chromatic aberrations. Although the mechanisms are not the same in both groups, the effect is the same. Lizards for instance have droplets of yellow oil in their retinal cones, while snakes actually have a yellow lens. The former also applies to turtles, which have red droplets, these being particularly effective in correcting chromatic aberrations.

In some lizards, including many gecko species, and all snakes the eyelids are absent, these being fused together to form a spectacle. In the crocodilians, chelonians and lizards that do have eyelids, the lower one is larger and movable. Added protection is given by a nictitating membrane, which is located in the corner of the eye and can be moved to cover and clean the eye, a function performed by the tongue in the lidless geckos.

Many lizards that live in arid environments have a transparent brille in the lower eyelid through which they can see if the eyelid is closed during a sand storm. This serves to protect the eye.

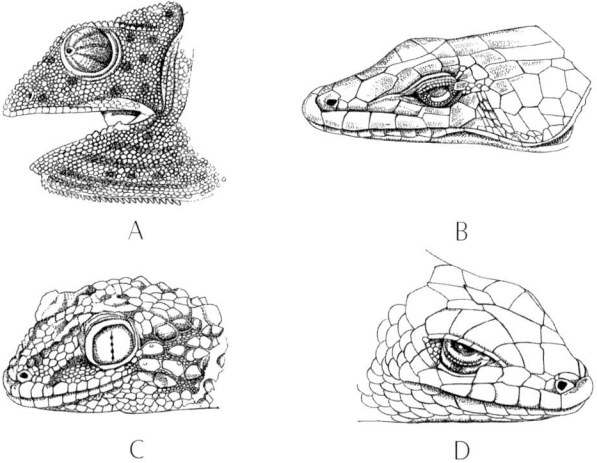

Various lizard eyes: A. Common chameleon, the eyes are particularly well adapted to its hunting methods, each working independently to locate prey and survey its surroundings; B. Certain terrestrial lizards have developed a brille in the lower eyelid, protecting them from windblown sand; C. The split eye of Turner's thick-toed gecko is typical of nocturnal animals. In bright light the pupil can narrow down to four pinpoint apertures; D. The eye of a typical diurnal lizard protected by an opaque eyelid.

Perhaps the most remarkable adaptation is encountered in the eyes of chameleons, which can rotate independently of each other, giving the animal an all round field of vision or 180 degrees per side. How the brain is capable of interpreting the two images perceived is not known. However, once one eye has detected prey the other eye also lines up as the head is turned, so that full bifocal vision of the prey is achieved and the tongue is flicked out to capture it.

SMELL

All reptiles, with the exception of the crocodile, possess vomero-nasal or Jacobson's organs although these are not well developed in turtles and chameleons. A familiar sight is the flickering tongue of snakes. This is in essence their smelling organ, as the forked tongue picks up scent molecules and transports them to pits or Jacobson's organ in the roof of the mouth. Sensory epithelia interpret these molecules and convey the results via the nervous system to the brain, which effects the necessary follow-up action. If a snake is in a new environment its tongue will flicker much more rapidly than in familiar surroundings. Lizards also flick their tongues though not as noticeable and mostly more slowly, although this varies according to species.

BRAIN

The brain of most reptiles is small and one would not consider them to be intelligent animals. Their behaviour tends to be stereotyped and mostly associated with food capture, reproduction and survival. In comparison to other vertebrates the brain shows many resemblances to that of birds. With the exception of crocodiles there appears to be little capacity for learning.

ALIMENTARY TRACT

The alimentary tract is basically the same in all reptiles. The mouth is mostly armed with peg-like teeth in lizards and sharp pointed in snakes and crocodilians, while tortoises, terrapins and turtles only have a horny beak. The teeth are not suitable as cutting instruments but rather serve to grip prey, which is then manoeuvred to the rear of the mouth and usually swallowed whole. Only tortoises and terrapins and some lizards hold the food down with their feet or claws, or use them to reduce it to a size that can be swallowed. The teeth are either mounted along the side of the jawbone (pleurodont) or directly along the ridge like that in mammals (acrodont). The teeth are relatively loosely attached and may break out but are soon replaced. Unlike mammals, the teeth in reptiles may be replaced throughout life. Teeth are not confined to the actual jawbones, that is the dentary, maxilla and premaxilla, but also occur on the pterygoid and palatine bones, especially in snakes.

The teeth in snakes are also differentiated into solid teeth without poison grooves (Aglypha) and teeth which have been enlarged with a poison groove down the front. In the back-fanged snakes (Opisthoglypha) the latter are located below or just behind the eye. In front-fanged snakes like the more primitive Proteroglyphs, the poison fang is located near the front

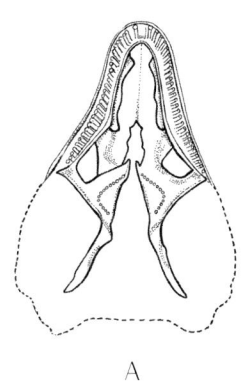

A

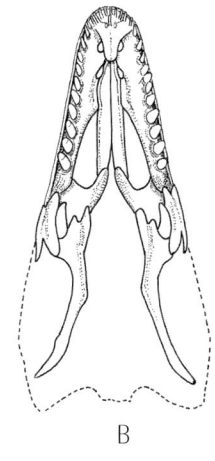

B

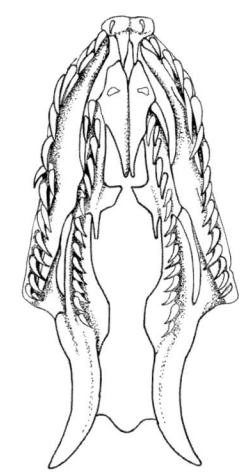

C

Different arrangements of reptile teeth: A. The Giant plated lizard, the teeth are pleurodont and are found on the premaxilla, maxilla and pterygoid bones; B. Water monitor, the teeth are also pleurodont, and are located only on the premaxilla and maxilla; C. Southern African python, the teeth are acrodontally arranged on the premaxilla and maxilla, pterygoid and palatine bones along the roof of the mouth.

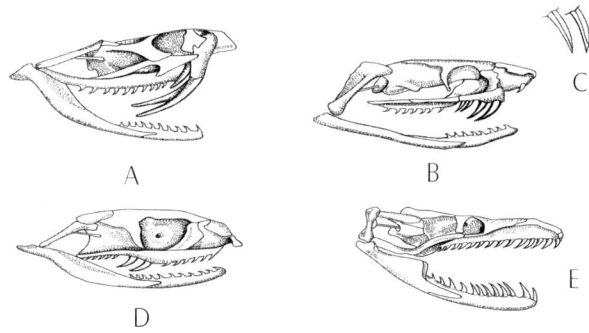

The different teeth arrangements in snakes: A. Solenoglypha, the poison fangs are attached to the moveable maxilla and as a result of their size are folded back against the roof of the mouth; B. Proteroglypha, the poison fangs are attached to the immoveable maxillary bone at the front of the snout; C. The poison fangs of the Proteroglypha can be split in two, those with a channel leading down the tooth to the tip and those in which the channel opens out on the front of the fang as in the spitting snakes; D. Opisthoglypha, the poison fangs lie just under or behind the eye; E. The Aglypha or solid toothed snakes, no poison fangs and all the teeth are the same.

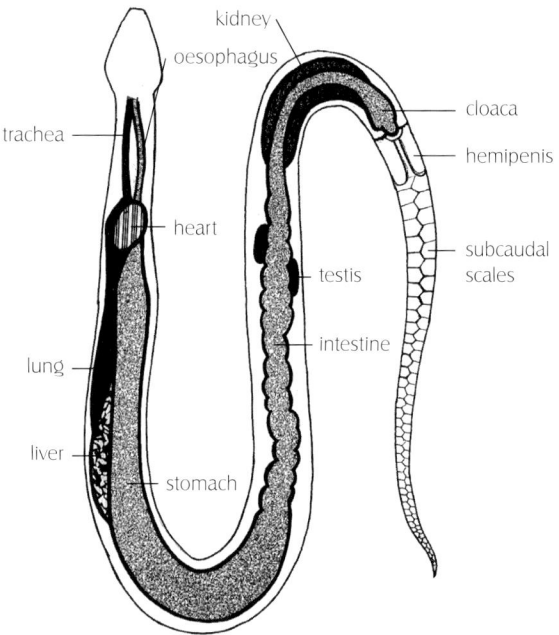

Schematic arrangement of the internal organs of a male snake. The narrow body has necessitated the reduction of the left lung, while the testes and the kidneys are unequally arranged alongside the intestines. The hemipenes lie inverted in the base of the tail.

of the snout and the groove is almost totally closed in front, forming a tube. The most advanced teeth are those of the adders (Solenoglypha), which have a hollow poison fang mounted on the moveable maxilla.

The short maxilla rotates like a hinge on the anterior end of the prefrontal bone. When at rest, the fangs are folded back and lie inside the mouth along the upper jaw. This is in contrast to the Proteroglyphs, which only have rigidly mounted fangs.

As mentioned earlier the teeth are mainly used to grip and hold the prey while it is crushed or manipulated for swallowing. Small prey may be swallowed in any way but larger prey is usually ingested head first, and enters the oesophagus, which is very distensible, passing down to the stomach by muscular action. The oesophagus is relatively long in reptiles and some species have modifications such as the oesophageal teeth of egg-eating snakes, formed by three elongated, downward projecting processes of the cervical vertebrae.

The stomach tends to be elongated and leads to the small intestine where, as is the case in mammals, digestion takes place, after which the food passes into the large intestine and rectum from where undigested waste in the form of faeces are voided through the cloaca, the latter also common to birds and amphibians.

In order to conserve moisture, which in the dry season is only available from metabolic activity and from prey, reptiles usually excrete the poisonous urea, a byproduct of digestion, as a concentrate of uric acid similar to birds and some amphibians. Water re-absorption takes place in the large intestine. If water is available most reptiles drink, usually by sucking it up. Some reptiles, especially lizards can store urine in a bladder, which opens up in the cloaca.

Respiration and circulatory system

Reptiles breathe through lungs, most species having two. Only snakes and some lizards have one lung on the right-hand side, or show various reductions in size on the left side. The lungs are not as well developed as in birds or mammals. Chameleons have sac-like structures similar to air sacs in birds enabling them to take up greater quantities of air. Breathing is similar to that in mammals although there is no muscular diaphragm, instead ventilation is brought about by muscular expansion and contraction of the thoracic cavity.

As a result reptiles are true terrestrial animals and therefore have a circulatory system to cope with this lifestyle. At the centre of this lies the heart, which in most species is three chambered but four in crocodilians. In most species the left and right auricles are separated, but the ventricle is undivided or partially divided. Full division of the latter occurs only in crocodilians, a condition similar to that in mammals.

THERMOREGULATION

Reptiles are mostly terrestrial animals and are cold-blooded or ectothermic being unable by physiological means to maintain their body temperature on a relatively even keel, and cannot function properly if this is not so. This varies according to species and from place to place. Animals living in temperate latitudes will not be subjected to the same temperatures as those from the tropics. Therefore the thermal tolerances of reptiles are dependent in part on their environment. As they do not have sweat glands, feathers or fur to regulate body temperature they must rely on ambient temperature and the sun to bring their body temperature to a degree at which normal functions can be carried out. This is important as at the one end they may be unable to move fast enough to catch prey and on the other they may die from overheating. The animals therefore bask during the early morning and once the right temperature is attained foraging and other activities begin. Each species has its specific temperature range, which for most tropical species exceeds 27 °C and may even reach 37 °C in some desert lizards. Prolonged exposure to the sun on a summer's day will kill a reptile, therefore relatively little activity takes place during the middle of the day, but is concentrated mainly in the morning and afternoon. Many species display quite elaborate sunning postures, usually flattening the body and lying at right angles to the sun's rays. Some snakes will kink the body while basking but to what purpose is unknown. Many species may be killed on tar roads when lying on the warm surface in the late afternoon and evening.

HIBERNATION

There is some dispute whether South African reptiles hibernate – which entails a shutdown of metabolic activity during winter, the animals lying dormant until temperatures rise – or whether they merely aestivate, that is lie dormant to wait out unfavourable conditions until these become more suited to activity. As most species do not feed during this time of the year, it can be construed that they do hibernate despite some localised activity by the animals. A few individuals of many species may be about on a warm winter's day, even on the highveld, often basking at the entrance to their retreat but little if any feeding takes place as temperatures fall too low during the night to allow digestion to take place. The inability to digest the food could kill the animal. Most reptiles rely on stored fat reserves to tide them over this period. The reptiles retire into holes or hollow logs, under logs and stones, hollow trees, under bark, in termitaria and in fact anywhere where there is protection from the cold. Hibernation in South Africa appears to be induced by day length and minimum temperatures, usually setting in towards the end of May followed by emergence during middle to late September.

During hibernation metabolic activity is markedly slowed down, especially the essential mechanism to control the use of stored fat reserves, so that the animal can last through the following three to four months.

During the dry and hot months of the year many species, especially tortoises may go into a period of rest or aestivation to avoid periods of food shortage and high temperatures. This is however usually of short duration mostly terminating following the advent of rain.

REPRODUCTION

South African reptiles breed mainly during the rainy season and mating mostly takes place soon after emergence from hibernation during September and October although a few species mate during late July and August. The main period coincides with the onset of the rains. There are many mating strategies, and courtship and rivalry are prominent. Displays are aimed at enticing females, while others act as deterrents to other males. However, a female may mate with more than one male and one male may also

The hemipenes of a Black mamba partially extruded showing the recurved spines which help to anchor the organ in the cloaca of the female.

serve more than one female. Males of some species such as flat lizards show numerous battle scars towards the end of the mating season. Fertilisation is internal and sperm is introduced into the cloaca of the female and transported via oviducts to the ovaries. Chelonians and crocodilians have a single penis but in lizards and snakes there is a pair of these organs known as hemipenes. These structures are characteristic of the Squamata and are of different shapes and sizes. Some are covered with spikes and knobs and their morphology is characteristic and consistent for a species. The different characters of hemipenal structure are important in preventing interspecific hybridisation and are therefore useful taxonomic aids.

In some species such as night adders sperm can be stored by females for lengthy periods, allowing successive clutches of eggs to be laid weeks or even months apart. However, the percentage of fertile eggs decreases with successive clutches.

Most reptiles tend to be oviparous or egg-laying and despite the risks appear to be highly successful. The eggs are laid in holes dug by the female or concealed under rocks or in other suitable sites, including the burrows of other animals. A female is likely to look at several sites before selecting one, where she proceeds to lay her eggs. They usually take about 10 weeks or longer to incubate and the emergence of the young mostly takes place in the mother's absence. A few reptiles – some snakes, a few lizards and crocodilians – show varying forms of parental care.

The eggs of tortoises and some turtles are hard-shelled, whereas those of most reptiles tend to be soft-shelled. Clutch size varies considerably, from one or two eggs in some tortoises to in excess of 100 in larger reptiles such as sea turtles and pythons. The egg is very yolky, providing for embryonic development and some may be retained after hatching to provide the hatchling with sustenance while it begins foraging.

Some snakes and lizards are live bearing or viviparous, the reptile embryos develop in the oviducts of the female, encased in a translucent egg membrane and virtually independent of the mother, being nourished by the egg yolk. In some species the young do have contact with the mother by means of a type of placenta, which may be important in assisting in the transport of water, oxygen and other gases from the mother to the embryo. The assistance is however supplementary to the yolk which remains the main source of food.

A few species are intermediate between oviparous and viviparous, or ovoviviparous, the young developing in the egg inside the oviduct of the female, so that once the eggs are laid the young hatch within the space of a few days. There is no contact between mother and embryo and this method merely represents a delayed egg-laying.

When the young hatch they cut through the egg membrane by means of an egg tooth situated on the tip of the snout, which falls off shortly thereafter. In those species where the eggs are buried the remaining fluid from the eggs moisten the soil enabling the hatchlings to dig their way to the surface.

While not being recorded from any South African species to date, the phenomenon of parthenogenesis has been recorded in both Europe and America. Some populations of lizards from the Caucuses mountains of eastern Europe consist almost entirely of females which reproduce by laying eggs which have not been fertilised by a male but which still develop normal lizards, mostly female but on rare occasions including functionally capable males. A recently introduced snake species to several coastal towns in South Africa, the Flower pot snake, *Ramphotyphlops braminus*, reproduces by parthenogenesis as is discussed later.

Growth and age

Most reptiles grow rapidly during their first few years of life provided that food is plentiful and temperatures remain at a relatively high level. This gradually decreases after the third to fifth year. However, growth may continue throughout life but decreases to very low levels later. In many lizard species females grow at a faster rate than males and sexual maturity is mostly reached earlier in females than in males. In many species of snakes, lizards and crocodiles males reach a larger size than females and in chelonians, such as the Leopard tortoise, the females are larger than males. Some reptiles only live for a year, hatching from eggs, feeding and growing, reproduce and after egg-laying they die. Other species may live for 20 years or longer and chelonians have been recorded reaching in excess of 100 years of age while crocodilians may reach 70 years or more. Most of our knowledge of longevity comes from captive individuals, and lifespan under natural conditions is likely to be considerably less due to predation pressure and environmental factors.

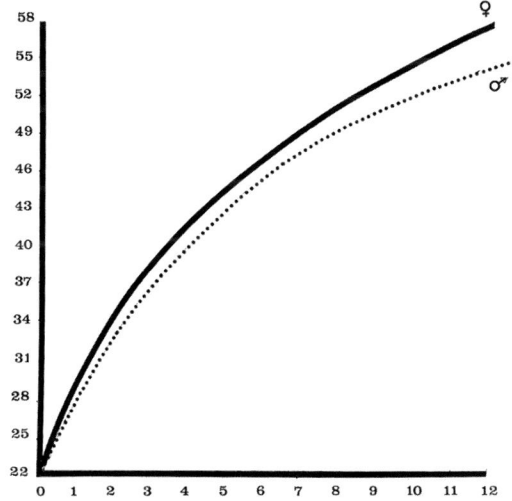

Differential growth between male and female in Variable skinks. The faster growth and larger size of the female are linked to the fact that she has to carry the eggs.

Zoogeography

The distribution of reptiles across various continents is very interesting, particularly when there are relationships between widely separated countries. The study of animal distributions is known as zoogeography. Some 200 million years ago the world formed a single universal continent called Pangaea. This gradually split up during the ensuing 20 million years to form a large northern landmass Laurasia and a southern equivalent Gondwanaland. At about this time India separated and drifted north, while Africa and South America together as one land mass began to drift away from Australia and Antarctica. About 135 million years ago Madagascar separated from Africa, but Australia and Antarctica were still joined. It is only during the last 65 million years that the continents have taken up their present position. India pressed up against the Asian continent, folding the hinterland, thereby forming the Himalayas ranges while Australia split off from Antarctica. The first fossil reptiles date back some 280 to 315 million years, reaching their peak in numbers and species about 200 million years ago, after which they declined and all but disappeared 70 to 80 million years ago when the era of mammals began, reaching its peak at the present time.

Only four of the 17 orders of reptiles have survived to the present day, which includes the tortoises, turtles and terrapins, crocodilians, snakes, lizards and the Tuatara. The turtles were already on the scene about 200 million years ago and had differentiated into two groups – the side-necked (Pleurodira) and the retracted-necked (Cryptodira) – about 20 million years later. The tuatara is the oldest survivor of the reptiles, fossils from that time having been recorded.

Australia, Africa and South America show remarkable similarities in some of their fauna such as fish, frogs, reptiles and birds, which all belong to similar families or genera. The theory of continental drift, widely accepted today, showed that there were connections in the form of land bridges with

Antarctica and between North and South America at a time when climates were less severe. Fossil records from Antarctica include species in common with both Africa and Australia. However, the long period of isolation of Australia has led to adaptive radiation of the dominant classes of vertebrates, such as the marsupials. Most of Australia's snakes belong to the cobra family Elapidae, which is also found in Africa and Asia, while being relatively rare in the Americas. The side-necked terrapins were common in the Americas, the Middle East and most of Asia some 135 million years ago, but today are restricted to two families, one from Africa, Madagascar and South America and the other from South America, Australia and New Guinea.

While the distribution of most of the present-day vertebrates can be explained in terms of continental drift and land bridges, the distribution pattern of reptiles is still changing today. Reptiles are hardy animals and can survive long periods without food or water. It frequently happens that during floods all sorts of debris and animals are swept down rivers and into the sea where currents transport them to distant shores. This method pertains particularly to small animals such as lizards. A small skink *Cryptoblepharus boutonii* or Bouton's snake-eyed skink recently colonised a rocky outcrop along the Zululand coast, having likely drifted south on driftwood or other debris from the nearest colony which is found near Inhambane about 700 km to the north. Many reptiles have been recorded 'hitching' on bunches of bananas and other produce and in crates on ships and by vehicle. Several species of lizard and snake are known to have colonised new areas in this manner. Some animals have, of course, been released inadvertently or on purpose in countries far from their original homes. Some species are extremely adaptive and settle down to thrive in an environment where there may be fewer diseases and predators. They can therefore compete with indigenous species for the same resources and threaten their existence. This is one of the main reasons why the authorities are against the importation of less desirable species. Some such widely known introductions into South Africa include the Red-eared terrapin, Indian myna, House sparrow, European starling, Black and Brown rats and many others. More local introductions include various strains of the Leopard tortoise, thereby risking the spread of diseases and affecting genetic purity of localised races.

These are some of the features, which make reptiles such remarkable animals. They have survived for millennia evolving into the great diversity now inhabiting the earth, almost from pole to pole, and in most environments. It is estimated that there are currently about 6 500 reptile species worldwide, but this number is changing daily as a result of surveys in hitherto remote areas as well as through the use of more advanced techniques for evaluating perceived differences and relationships.

South Africa is blessed with an exceptional array of reptiles with 363 species having been recorded, about 6 per cent of all species occurring in the world, and that on only about 1 to 2 per cent of the earth's land surface area. Of these 177 (or 48 per cent) are endemic or near endemic to South Africa, that is they occur either totally or mostly within the borders of the country (some extending marginally into Namibia or Zimbabwe) and nowhere else in the world. The number of species recorded is still far from absolute as new species are constantly being discovered, with advances in technology and new insight. However, this richness is being threatened by a burgeoning human population and ever-increasing demands on the land and its resources.

A Veld monitor lying motionless in a rock crevice to avoid detection.

The Geometric tortoise is an attractive species, which accounts for the large numbers exported for the pet trade in earlier times. Photo: Richard Boycott

Tortoises, Terrapins and Turtles

Order Chelonia

Tortoises have always been considered endearing animals by the majority of people, even those who find other reptiles repulsive. In spite of being well known, tortoises and their relatives are highly unusual animals, both in behaviour as well as anatomy, which distinguish them from all other reptiles. In spite of considerable confusion with regard to common names, tortoises, turtles and terrapins all belong to one order Chelonia. Tortoises have remained unchanged for longer than most groups of animals. Although the anapsid line of the chelonians originated some 200 million years ago, modern tortoises appeared and have changed little in the last 55 million years. Several fossil tortoises have been described from South Africa, including species of Psammobates, Homopus, Chersina *and* Geochelone *ranging in age from the Miocene 25 million years ago to the Holocene about 10 000 years ago. Today there are 273 species of chelonians living worldwide, representing about 90 genera in 25 families, of which 13 genera and 24 species (including an introduced species) occur in South Africa and in the waters off her coast, including both the largest and the smallest species.*

Skeleton

The most characteristic feature of a chelonian is its shell, which consists of a bony skeleton covered by a thin horny layer or series of plates, the lamina, which lie across and cover the sutures between the bony plates in the underlying skeleton. The shell is divided into two sections, namely the carapace above and the plastron below, the two being connected by bridges between the front and hind legs on each side of the shell. The arrangement and number of the lamina and the arrangement of the underlying bony plates are important guides to the classification of these animals. The lamina correspond to the epidermis of ordinary reptilian scales and are produced by living cells which lie between the lamina and the bony plates beneath. As the animal increases in size the older lamina increase by the addition of new material along the margins. Tortoiseshell, used to cover brushes, mirrors and other cosmetic jewellery, is lamina derived from the shell of the Hawksbill turtle, *Eretmochelys imbricata*, when the animal is killed or when the lamina are induced to part from the bony carapace beneath by slowly roasting the animal alive until the heat causes the lamina to lift off the underlying bone. It is claimed that the lamina subsequently regenerate but this is doubtful and the animal more than likely dies.

The shell of most species is rigid with the exception of the hinged terrapins, genus *Pelusios*, where there is flexure between the pectoral and abdominal shields, and the hinged tortoises, genus *Kinixys*, where the flexure is at the rear of the carapace in front of the hind legs. The retractor muscles responsible for closing the shell are exceptionally strong. Both methods are invaluable defence mechanisms.

Other remarkable modifications include a complete reduction in the number of vertebrae, as the rigid shell removes the need for a flexible backbone. The shell protects the vital organs and assists in their support. There are only 12 individual bones apart from those of the neck and tail. The vertebrae are mostly attached to the underside of the bony plates of the carapace. Most of the ribs are also fused with the carapace and partially merge with the shell. Due to this rigidity, the back muscles have also become much reduced. However, in the neck and tail region the spine is flexible and the surfaces between the neck vertebrae are highly modified and allow kinking when the head is drawn back into the shell. Of further interest is the fact that the pectoral and pelvic girdles to which the limbs are attached lie inside the rib cage and not outside as in other vertebrates. The sternum or breastbone, the collarbone or clavicle and another small bone present in reptiles, called the interclavicle, are entirely absent. The rigid structure of the carapace makes their use superfluous and they have fused with the plastron.

The rigidity of the shell serves as protection, with gaps only present at the front and rear. This rigidity presents the animal with other problems, such as how does the animal breathe. The belly muscles have been modified to form part of a specialised breathing apparatus quite different from that of other vertebrates. One set of muscles widens the body cavity so that the

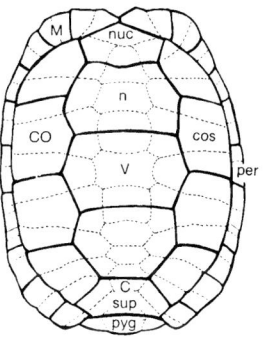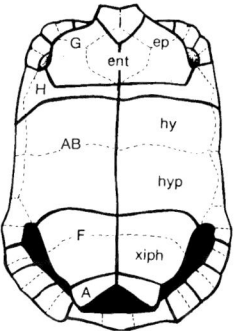

Epidermal lamina (—) and underlying bone formation (···) of a Leopard tortoise shell.

M – marginal; CO – costal; V – vertebral; C – supracaudal; I – intergular; G – gular, H – pectoral; AB – abdominal; F – femoral; A – anal.

Shell bones: nuc – nuchal; n – neural; sup – suprapygial; pyg – pygial; cos – costal; ep – epiplastron; ent – entoplastron; hy – hyoplastron; hyp – hypoplastron; xiph – xiphiplastron.

The number and arrangement of lamina and underlying bones vary between species and are an aid in their identification.

A chelonian shell, the lamina overlap across the underlying bone sutures, covering the skeleton.

lungs can fill with air while another set forces the viscera against the lungs causing the air to be expelled. Some freshwater chelonians can suck water into the mouth and pass it over membranes in the throat, acting as a type of gill. Others have specialised thin-walled sacs opening off the cloaca, which may function in a similar manner.

The chelonian skull is unlike that of other modern reptiles, and with the exception of some of the oldest fossils, has no teeth, the mouth being framed by a hard horny beak similar to that of birds. As a result of this the animals have to bite off pieces of food and swallow it whole. Tortoises, which feed on hard food, have developed specialised flat or ridged crushing surfaces on the roof of the mouth and large strong jaws. Many species, especially terrapins, frequently use their forefeet to assist in tearing the food into manageable pieces.

Food

The diet of chelonians is variable, land tortoises are mainly vegetarian although they will also take animal matter. Terrapins or freshwater turtles are largely carnivorous and feed on invertebrates, fish, amphibians and even the young of aquatic birds and mammals. The large sea turtles live on a diet of plant and animal food. Although chelonians can go for lengthy periods without food, they feed frequently and can build up fat reserves. Food appears to be digested slowly and is dependent on ambient temperature. Tests done on one species showed that the rate of digestion more than doubles with a rise in temperature from 18 °C to 29 °C.

Hibernation

During winter the animals go into a state of hibernation or prolonged inactivity during which all metabolic processes, including growth, appears to be retarded. The dormant period is marked by the formation of a ring around previously formed material on each lamina. In species that hibernate regularly these rings are added at the rate of one a year, but this does not appear to apply to South African conditions. Such rings on local species are not readily assessed and the method is unreliable. Disease or food shortages could produce similar rings, and wear and tear on older laminae obscure them.

Sexual dimorphism and reproduction

There is little external sexual dimorphism among most chelonians. Females tend to be larger than males, and the male may possess a concavity in the plastron to assist in holding him in place during mating. Some chelonians have enlarged gular processes in males, such as that of the Angulate or Plowshare tortoise, *Chersina angulata*, which assists the male in fighting other males over a female. The tails of males are usually longer than that of females.

Courtship varies between species and includes butting or biting of the female by the male. Nodding of the head under these circumstances is a common phenomenon. A more elaborate display involves stroking the female's head with elongated claws of the forearm. Aquatic chelonians mate in the water. When copulating the male mounts the female and inserts his penis under the shell into the cloaca, resting his forelimbs on top of the shell and holding his body in a semi-upright position. Mating can be noisy and during the process the male utters a variety of sounds, from a roar to a high-pitched scream or simply wheezing.

Although mating usually takes place prior to egg-laying, this is not always the case as the female of some species can retain viable sperm and fertilise her eggs for up to four years. This adaptation is also known among the snakes and may be present in lizards as well.

All chelonians lay eggs, which in the case of land tortoises and some terrapins have hard shells, while other terrapins and turtles lay soft-shelled eggs. The eggs are round to oval and number from one to two hundred according to the species. They are laid in a hole in the ground dug by the female. Sea turtles nest at night whereas other species are diurnal. More than one site may be tried before one is selected. The eggs are well covered and the soil compacted. They hatch after two to twelve months depending on the species and time of the year that the eggs are laid.

ENEMIES

Chelonians have many enemies, including birds such as herons, storks, eagles and ground hornbills, reptiles such as monitor lizards and crocodiles, and mammals. Certain ticks feed primarily on tortoise blood and may congregate in large numbers in exposed areas around the axis of the limbs and around the cloaca and tail. Humans in particular destroy many animals for food, pets and when burning the veld.

During winter when most tortoise species are hibernating, many take shelter under rocks, logs, grass tussocks or piles of dry vegetation. It is this habit of sheltering under logs and vegetation that leads to large numbers being killed in veld fires. Those that survive are often badly scarred, and in many cases the laminae of the carapace are totally burnt off exposing the bone. As there is no regeneration of these laminae the bone is permanently twisted and scarred and it is incredible that some of these animals survive. Domestic animals such as dogs also contribute to the carnage, adding to the threat facing these unique creatures.

Chelonians have a long history and association with humans. In China they are considered to represent one of four spiritual animals responsible for longevity and a symbol of righteousness. Tortoises carved out of stone are found at temples throughout the country. Even the Greek gods regarded the tortoise as holy and Pan, the god of the mountains and shepherds, was its protector. Various North American Indian tribes honoured the tortoise and made ceremonial rattles from their shells, while the Bushmen of the Kalahari used the shells to carry snuff in.

Most chelonians are eaten by humans, which present the greatest threat to the survival of most species. The giant tortoises of the Galapagos Islands and the Seychelles were used for meat aboard the sailing ships of Spain, Portugal, Holland and England. The word 'Galapagos' is derived from the Spanish for tortoise.

Sad to relate, these tortoises were captured because they live for several months without food or water and were therefore a source of fresh meat for the sailors on the old sailing ships. This led to the eventual extermination of several forms on the Seychelles prior to 1800. Only one subspecies of the giant tortoises was left on the island of Aldabra, which because of the sheer cliffs and rocky shores was relatively inaccessible. Despite this it was recorded that 1 200 of these tortoises had been taken off this island in 1847. In the Galapagos Islands the tortoises fared better and only a few subspecies became extinct.

Turtles appear to be as important as a source of food, probably because of their size and their widespread occurrence. They come to land in large numbers in order to lay their eggs and are then easily harvested. Most chelonians are still eaten today in countries around the world, but especially in China, Japan, Cambodia and the United States. This applies especially to the soft-shelled terrapins, while the eggs of the sea turtles are also used worldwide.

The Romans used tortoiseshell to decorate furniture. It is still used today in jewellery, combs and brushes, hair ornaments and jewellery boxes. There are various other uses, including high-grade machine oil derived from the fat of some species. Chelonians are much sought after in the pet trade, which ranks as one of the greatest threats to these animals.

The Order Chelonia can be divided into the Cryptodira or retractable-necked species and the Pleurodira or side-necked animals. In South Africa all the sea turtles and land tortoises belong to the former while all freshwater terrapins belong to the latter. There are 13 tortoise species, four terrapins and five turtles, which occur on South African soil and in its territorial waters. This is an extremely rich heritage and ranks among the highest in the world. In terms of land tortoises South Africa boasts the highest diversity, for which we should be proud. A fifth terrapin species has been introduced into the country as pets, and has managed to establish itself in the wild in some urban areas.

Side- and Snake-necked Terrapins
Suborder Pleurodira

The Side-necked Terrapin Family
Pelomedusidae

The side-necked terrapins form two families, both restricted to the southern continents, but only one, the Pelomedusidae, occurs in South Africa. Although this family is more tropical in distribution, five species are found here. The side-necked terrapins are characterised by being able to fold their necks sideways into the anterior orifice of the carapace. This is made possible by flexure joints just behind the head, in the middle of the neck, and where the neck vertebrae adjoin the first trunk or thoracic vertebrae. This enables the neck to be completely retracted into the shell.

The most common and well-known species is the Cape terrapin, *Pelomedusa subrufa*, which occurs in South Africa from the Cape Flats to the Limpopo River in the north but is widespread in sub-Saharan Africa. In some areas such as the North West Province, it is very abundant and as many as 16 individuals have been recorded from a single small depression filled with water. They are plentiful in dams and rivers during the rainy season but leave the water from the end of May until August during which time they lie buried in the soil some distance from the water. They also do so if the water in a pan dries up, but emerge again as soon as sufficient rain has fallen to soak the ground.

Terrapins are highly aquatic animals but breathe by coming up to the surface. They are commonly seen extending their heads to just above the water or just pushing the nostrils above the water surface, but will submerge again if alarmed. They are very curious animals and will survey an intruder from this position. They can be seen early in the morning basking on rocks or logs in a dam. The head is held high as they lie soaking up the sunshine. However, if disturbed they will slip off sideways or simply dive into the water. Their eyesight is good and they are very alert. They can remain under water for long periods when disturbed but being curious will rise to see what is happening.

These terrapins feed during summer when they are most active, and are mainly carnivorous. They eat fish, frogs, tadpoles and fledglings of waterbirds and have been recorded ambushing doves as the birds come to

The Cape terrapin is typical of standing bodies of water.
Photo: W.D. Haacke

The Serrated hinged terrapin is mostly found along rivers in the northern and eastern parts of South Africa.

drink, biting a foot and dragging the bird under and drowning it. They are strong animals and it is said that they will even take ducklings. When feeding they bite into their prey and with forward thrusts of the forefeet rip off pieces which are then swallowed. This process is repeated until the prey has been consumed or the animal is satisfied.

Mating takes place in summer and the eggs are laid during November and December in a hole in the ground. This is dug by the female to a depth of 10 to 15 cm and a width of about 7 cm. If the soil is hard the female urinates on it and then excavates the softened earth. The excavated soil surrounds the hole and is later used to seal it. When the hole is of the right dimensions the female lays up to 16 oval soft-shelled eggs and then scrapes the soil over the eggs. The moist soil is packed hard by the stamping and levelling action of the female until all signs of the nest are obliterated.

These terrapins are renowned for the foul odour they exude. This is produced by secretions from glands in the skin between the plastron and cloaca and probably assists in deterring predators. They frequently exude this malodorous fluid if molested, which indicates that this is a defence mechanism. It is also a contact mechanism between male and female or a recognition signal among members of a pond.

The four hinged terrapins are restricted to the larger east-flowing rivers, swamps and pans of Limpopo, Mpumalanga and KwaZulu-Natal provinces. They are characterised by the hinge at the front of the plastron, which closes the gap when the head has been retracted. Both the Mashona hinged terrapin, *P. rhodesianus*, and the Yellow-bellied hinged terrapin, *P. castaneus*, appear to have very limited distributions in South Africa, having only been recorded along the coast of KwaZulu-Natal. Similarly, the Pan hinged terrapin, *P. subniger*, is also very rare in South Africa, having only been found in pans in the Wambiya sandveld of the north-eastern Kruger National Park. It is however likely to occur in suitable habitat along the Limpopo River as it has fairly recently been found to occur along the Motloutse River in eastern Botswana.

The most common of the four species is the Serrated hinged terrapin, *P. sinuatus*, which occurs commonly along rivers in the lowveld of Limpopo and Mpumalanga provinces as well as those of northern Zululand. It is frequently seen in the Kruger National Park basking on rocks and logs in rivers. All of these terrapins are primarily carnivorous, feeding on invertebrates and small vertebrates as well as scavenging meat from carcasses of animals that have died in the water or have been killed by crocodiles. Despite being eaten by crocodiles, the Serrated hinged terrapin has been seen feeding amongst crocodiles at a carcass.

Although they bite and scratch with their horny beak and claws, their chief defence appears to be the production of a foul-smelling fluid from glands in the cloaca, which may act as a deterrent to terrestrial predators.

HIDDEN-NECK TORTOISES AND TURTLES
SUBORDER CRYPTODIRA

THE LAND TORTOISE FAMILY
TESTUDINIDAE

This family contains all of the terrestrial tortoises, which are mostly characterised by a high carapace, and range in size from species with a length of 60 mm to the giants of the Seychelles and Galapagos Islands, which reach a length of 1 200 mm with a mass of 254 kg. This is, however, still small when compared to the extinct *Colossochelys atlas*, which had a shell length of 1 800 mm. This animal lumbered across the hills of southern Asia about a million years ago.

Most widespread in South Africa is the Mountain or Leopard tortoise, *Geochelone pardalis*, so-called because of its size and colour pattern. The eastern Cape forms are particularly large and imposing, reaching a length of

A Leopard tortoise drinking water.

Leopard tortoises mating.

700 mm and a mass of 40 kg. It is thought that this may be due to their diet as the species only reaches half this size in the northern provinces of South Africa.

The Leopard tortoise is a vegetarian and is active only during the summer months in South Africa when food is optimal and the animal can build up food reserves to tide it over the long, dry winter. Tortoises do not have teeth but with the aid of a horny beak they shear through succulent grass and herbs. Only small bites are taken and swallowed whole. On rare occasions dead animal matter may be consumed, while they are known to chew on bones apparently for calcium, which is needed for bone and shell development as well as for egg formation.

Towards the end of May the animals begin to search for a place to 'hibernate' or overwinter, a tendency that is well known. There is a story of a farmer who picked up a tortoise in the veld and placed it in a bag, planning to release it at home in the garden. On reaching the shed he hung the bag on a nail and completely forgot about the unfortunate animal until quite by chance he took the bag down again a year later. He was amazed to find the tortoise alive and apparently none the worse for its experience. He promptly released the animal and it disappeared into the bush.

Places to hibernate are varied, such as holes in the ground or sites among rocks or under logs, where the animal digs its way in or under, with only the shell protruding. Reptiles are able during hibernation to slow down their metabolic activity, thereby reducing the breakdown of stored food to a minimum. This enables them to survive for long periods without food or water. In fact, tortoises can go without water indefinitely as green plants have a high liquid content. However, if water is available they drink readily, especially in captivity.

Mating takes place in September when the weather starts warming up and animals become active once again. About a month later the female will look for a nest site and then proceed to dig a hole with her hind feet. If the soil is too hard she may urinate on the spot after which the soil is scraped out. This alternate wetting and digging continues until the hole is of the right depth. The size and depth of the hole varies in relation to the size of the female and can be as much as 15 cm in diameter and 20 cm deep. The hole is bottle-shaped with a bulge in the middle. Depending on the size of the female, up to 24 eggs are laid in the hole. These vary in size from 30 to 50 mm and are spherical and hard shelled. After the eggs have been laid the soil is replaced and stamped down using the feet and plastron, then the eggs are left to incubate. Sometimes the nest is camouflaged with bits of grass and scraped up leaves.

Within a day of depositing her eggs the female can mate again and the process is repeated. Several clutches may be laid during the season until environmental conditions bring this to a halt. It is interesting to note that in captivity mating and egg-laying can continue virtually throughout the year because environmental factors such as food shortages and temperature extremes are not severe enough to limit activity. After mating the male loses all interest in the female and lumbers off.

The eggs stay in the ground for variable lengths of time depending on the climatic conditions and when the eggs were laid, so that incubation may take anything between four and 15 months. The young hatch and make their way to the surface. They are able to dig their way out as the ground has been moistened by egg residues. On hatching they are about the size of a R5 coin and their shells are very pliable at this stage. The bones gradually grow together and the sutures meet so that after two to three years complete ossification has taken place. Tortoises grow relatively slowly but this varies from area to area and correlates with climatic and food variability. A four-year-old individual under captive conditions may reach a plastron length of 20 cm or more.

Leopard tortoises are relatively long-lived animals and under normal conditions can reach an age of up to 100 years. The Giant tortoises, *G. gigantea*, of the island Aldabra in the Indian Ocean have been recorded reaching 152 years of age. It is possible that they exceed this age, but it is difficult to obtain reliable records over longer periods of time.

The Leopard tortoise lives in a distinct home range to which it will return even if transported several kilometres away. This homing instinct has been the subject of many stories, the most famous being that of a tortoise picked up

The Geometric tortoise, an endangered species as most of its habitat has been destroyed. Photo: Richard Boycott

near Beaufort West and taken to Marico in Northwest Province. According to legend this tortoise returned to its place of origin, some 800 km away, taking three years to complete the journey. This is highly unlikely, but it is not uncommon for animals to travel distances of between 1 and 5 km. Home ranges vary according to availability of food and shelter as well as the sex of the animal, with females having larger home ranges than males. In the Addo National Park an average home range of 57 ha was recorded while that of a female in the Serengeti in Tanzania was estimated to be about 150 ha.

The Leopard tortoise is well known to South Africans and is kept as a pet throughout the country. Unfortunately this results in animals being transported from the Cape provinces to the northern parts of the country where they hybridise with local populations, as many escape or are released by owners who have grown tired of them. This is unfortunate as genetic strains mix and the animals lose their identity, besides the danger of introducing diseases to local populations to which they may succumb and be exterminated. Keeping tortoises as pets is a thoughtless act and should be discouraged.

The remaining 12 tortoise species are much smaller and extremely interesting but less well known as some, in particular the Geometric, Tent and Kalahari serrated tortoises of the genus *Psammobates*, are difficult to maintain in captivity. They are mainly inhabitants of arid areas where they feed on specific plants including succulents.

The Geometric tortoise or Suurpootjie, *P. geometricus*, is one of the world's rarest tortoises and endemic to the fynbos of the south-western Cape. Extensive surveys of remaining natural habitat have shown that this species now only occurs on tiny fragments of land. Only a few hundred survive in five provincial nature reserves with the majority occurring on private farms, in particular the Elandsberg Private Nature Reserve which houses the largest number, an estimated 2 700 to 3 400 individuals. Densities range between 0,6 to five individuals per hectare, depending on the quality, extent and condition of the habitat. An attempt at reintroduction onto a private nature reserve appears to have failed as that population has declined to 25 per cent of the original number introduced.

These tortoises are active throughout the year. During the day, activity takes place mostly during mid-morning and mid-afternoon, according to climatic conditions, the animals keeping their body temperature three to six degrees above the ambient temperature. Like all tortoises, their principal diet consists of forbs and grasses, but they will ingest some animal matter as well.

They breed during spring and early summer when a single clutch of two to four eggs is laid. The eggs measure about 32 mm × 24 mm and are laid in a hole in the ground dug by the female. Incubation takes between five and seven months, and hatchlings are found from March to May. Growth appears to be rapid, males growing at more than twice the rate of females and individuals reach sexual maturity at about seven to eight years. It is estimated that individuals may attain an age of 10 to 15 years.

During the last century large numbers of this beautiful tortoise were exported overseas, ending up in pet shops. Because of their relatively specialised diet few survived more than a year. It was only as recently as 1951 that the export of these animals was prohibited, but by then there were precious few left. To compound matters, many people, both black and white, regard the Suurpootjie as a delicacy and the unfortunate animal is cooked by dropping it in a pot of boiling water.

Today these tortoises are internationally recognised as an endangered species and legislation exists which protects them from possible exploitation. Despite this the greatest threat comes from habitat destruction. It is estimated that up to 96 per cent of their original habitat has been destroyed, mostly by agriculture but also housing and other forms of degradation. Uncontrolled fires have decimated many populations and remain a major threat to their continued survival. Yet it is necessary from time to time to burn parts of their habitat in order to promote the growth of specific food plants required for the maintenance of the Geometric tortoise population.

The Tent or Knoppiesdop tortoise, *P. tentorius*, is particularly beautiful with the carapace formed of large conical knobs covered by shields, each shield being black above with bright yellow to orange rays radiating from the centre to the margin. There are three subspecies, all of which are endemic to South Africa although one extends into southern Namibia. They are small tortoises, the length of the plastron ranging from 8 to 12 cm. They are less active than either of the previous species discussed and inhabit dry karroid areas, feeding on succulents and other plants. They lay two to three eggs in early September and hatchlings, which measure about 25 mm in diameter, have been found in May.

A similar but more widespread species is the Kalahari serrated tortoise, *P. oculiferus*, which occurs in a wide band from eastern Namibia through the Kalahari to the Northern Cape, western Free State, Northwest and Limpopo provinces in South Africa. The species is so named because the posterior marginal scutes curve upwards and are pointed, forming a serrated edge to the shell. This species is found as far east as Sekhukuniland. Here remnant areas of windblown or aeolian sand are still present, a reminder of arid periods during the Pleistocene, about a million years ago when such wind-borne materials arising in the Kalahari were deposited over a wide area in northern and western South Africa. The shells of this tortoise are commonly used by the Bushmen of the Kalahari as snuffboxes and containers.

Another remarkable species is the Ploughshare or Angulate tortoise, *Chersina angulata*, a common species along the coast in the Eastern and Western Cape provinces. The male is armed with an elongated protrusion of the plastron, which extends far forward and supports the head. This protrusion is used specifically to combat other males, particularly during the mating season. In females it is less pronounced, but still much more developed than in other species. The dorsal colour is yellow with a broad black margin and often with black in the centre of each shield. The belly or plastron is red-brown with a black centre. It is this brick-red colour that gives the species the common name Rooipens or 'red belly'. When a female is nearby, a male will run up to her and sometimes turn her over in his eagerness to attract her attention. In fact, he may turn over any other tortoise in the vicinity. Should he meet a rival male, a battle ensues as each one endeavours to position his gular shield under the opponent to roll him over. Sometimes, while in this

TOP: *One of the subspecies of the Tent tortoise,* Psammobates tentorius trimeni, *endemic to Namaqualand and extreme southern Namibia, is the most attractive of South African tortoises. Photo: Richard Boycott*

CENTRE: *A Kalahari serrated tortoise, widespread in the arid areas of South Africa.*

BOTTOM: *The Ploughshare tortoise, a widespread and common coastal species.*

The endemic Southern speckled padloper, an inhabitant of arid rocky areas in the western parts of the Great Karoo, is one of the world's smallest tortoises. Photo: Richard Boycott

The Southern speckled padloper, emerging from a rock crevice, its refuge during the heat of the day. Photo: Richard Boycott

fighting mood, the winning male will continue to push the overturned rival around before walking off to court the female, following her while nodding his head rapidly. The conquered male is left struggling to right himself. Should he manage to find some purchase he may be able to right himself, but if he is unsuccessful he could die from exposure.

The female lays a single egg at a time and this appears to be repeated throughout the season. The egg is oval to round and approximately 25 mm in diameter. It is laid in a shallow hole about 10 cm deep.

Although widespread and occurring in substantial numbers, habitat destruction through agriculture, urbanisation, the spread of alien vegetation and the indiscriminate use of fire has decimated populations. Counts of tortoise mortality at two different sites following fires found that between 54 and 77 per cent of the tortoises had succumbed – a truly alarming statistic.

The padlopers form a group of four species, which are endemic to South Africa. They are mostly restricted to the arid parts of the country and do not survive long in areas where there is considerable humidity. Two of the species, namely the Karoo padloper, *Homopus boulengeri*, and the Speckled padloper, *H. signatus*, are relatively rare and not well known. Both tend to inhabit rocky terrain taking refuge under stones and in crevices between rocks, often on rocky outcrops and hills. The Speckled padloper is regarded as the world's smallest tortoise, adults having a length of only 60 to 80 mm. With the exception of the Lesser padloper or Parrot-beak tortoise, *H. areolatus*, which has dark centred, concave dorsal shields and is very attractive, the other species tend to be red-brown, brown or olive-brown. The Greater padloper, *H. femoralis*, is the largest of the four species with a carapace reaching 14 cm in length and a mass of 700 g while the Lesser padloper may weigh up to 250 g. An amateur herpetologist W. Archer, who studied this tortoise, observed that it may feed so well in captivity that it outgrows its carapace, resulting in an inability to withdraw into its shell. Either the hind parts or the head and foreparts protrude depending on which side is first withdrawn into the shell. Both the Greater and Lesser padloper are egg-laying, normally producing two eggs. The hatchlings are tiny and very soft, as bone formation takes place gradually, although it is already evident at four months.

The last group of land tortoises are the hinged tortoises, of which four species are found in South Africa, one of which is almost endemic, the other three species being found beyond our borders as well. The hinged tortoises occur in the northern provinces and KwaZulu-Natal, but are absent from southern KwaZulu-Natal and the highveld regions of Gauteng and Mpumalanga provinces. The Lobatse hinged tortoise, *Kinixys lobatsiana*, is restricted to the bushveld of North West and Limpopo provinces, marginally crossing into eastern Botswana. Its distribution appears to overlap along its northern and eastern limits with that of Speke's hinged tortoise, *K. spekii*, while the Natal hinged tortoise, *K. natalensis*, occurs in northern KwaZulu-Natal and the southern lowveld of Mpumalanga as far north as the Olifants River. The fourth species, *K. belliana*, is limited to coastal Zululand and Maputaland, extending across the border into Mozambique.

These relatively nondescript tortoises are unusual in that they have a point of flexure in the carapace on either side in front of the hind legs, which is most visible in adults. This flexible joint results from the arrangement of the side bone plates, which have not joined along the sutures. The plates are held in place by a very strong skin, acting as a hinge, enabling all members of this genus to close up the hindquarters after the legs are withdrawn into the carapace. The foreparts, including the head, are protected by the heavily scaled forelimbs, which can be withdrawn into the carapace and block any opening. It is interesting to note that the hinge only becomes visible in the adult animal. During the juvenile stage the plates are flexible and can bend but there is no sign of the hinge.

These tortoises are found most frequently in rocky areas, but not exclusively so and are just as much at home in bushveld country. They hibernate and aestivate under rocks and in fissures, where they are well protected, but many go under logs and even piles of brush. This is very dangerous as winter is the time of uncontrolled veld fires and many are incinerated before they can escape. Apart from fire they have many enemies, especially juveniles and smaller individuals, and are eaten by dogs, jackals, smaller carnivores, crows and ground hornbills.

Another unusual aspect of these tortoises is their diet. Apart from feeding on typical tortoise food such as grass, herbs and flowers they frequently include animals in their diet. This may be in the form of dead animals which they scavenge or live animals such as millipedes. They are also known to chew on bone to obtain calcium, which may be lacking in their diet.

These tortoises lay their eggs during summer, usually up to five at a time. The soft-shelled hatchlings appear some 10 months later and measure approximately 40 mm in length and almost as much in

A Greater padloper, widely distributed in the Northern Cape and part of North West Province.

The Lobatse hinged tortoise, a near endemic tortoise of the bushveld of North West and Limpopo provinces.

width, and are vulnerable to attack by ants at this stage. Growth is slow at first, but increases more rapidly as the animals get older, slowing down again later. Adult specimens rarely exceed 20 cm in length and 14 cm in width, with a mass of between 1 and 1,5 kg, although frequently smaller.

In contrast to hibernation, which is brought about by a shortening of day length as well as a change in temperature, these tortoises and most other species undergo a similar but less extensive period of aestivation in summer to escape the heat, especially during hot, dry periods. They seek shelter and keep out of direct sunlight, but as soon as rain has fallen they become active and can be found roaming about, feeding or following a mate. In some places the local people are able to predict changes in the weather pattern by observing the behaviour of these animals.

THE POND TERRAPIN FAMILY

EMYDIDAE

The American red-eared terrapin may be established in the PWV area of Gauteng and in Durban. It poses a threat to indigenous species.

These are commonly called 'typical terrapins' but include some bizarre forms such as the Big-headed terrapin of China and south-east Asia. Most are, however, 'typical' terrapins and include such brightly coloured species as the American painted terrapins of the genus *Trachemys* one of which, the American red-eared terrapin, *T. scripta*, has escaped from captivity in South Africa and elsewhere.

This is a very attractive terrapin when young but becomes less colourful with age. Animals have escaped and been released into streams in the Pretoria, Johannesburg and Durban areas. Tolerant of a wide range of environmental conditions they have as yet not become well established in South Africa but have the potential to do so. They live in quiet backwaters and pools and hibernate when minimum temperatures drop below 10 °C.

THE SOFT-SHELLED TERRAPIN FAMILY

TRIONYCHIDAE

This unique family of terrapins is characterised by a soft shell or carapace. Approximately 30 species are known worldwide. Only one, the Zambezi soft-shelled terrapin, *Cycloderma frenatum*, has been recorded as far south as the swamps and pans along the Sabi or Save River in southern Zimbabwe and Mozambique. In addition to its peculiar carapace this animal has nostrils at the end of a snorkel-like projection. Another species is the Nile soft-shelled terrapin, *Trionyx triunguis*, which has an even more pronounced drawn-out nose terminating in the nostrils. It has been recorded from the mouth of the Cunene River on the border between Namibia and Angola.

Sea turtles
Suborder Cryptodira

Having briefly dealt with the tortoises and terrapins, a discussion of the chelonians would not be complete without mention of the giants of the ocean, the sea turtles, which are found along the east and to a lesser extent the west coasts of South Africa. They are magnificent animals and five species are known to visit our shores, but only two of them nest along the Maputaland coast. The five species include the Leatherback, Green, Loggerhead, Hawksbill and Olive Ridley turtles. They belong to the same suborder as the tortoises because of the way in which the head is adapted to be withdrawn into the carapace.

Turtles spend their whole life in the sea – with the exception of when they come ashore to lay their eggs – floating, swimming and drifting with the currents, often appearing far from their main haunts. Most of the time these turtles are found in warm waters but individuals, especially juveniles, may occur as far south as Cape Agulhas. They spend up to three years drifting in the ocean, feeding on floating organisms, such as bluebottles and storm snails. Following the ocean currents they are eventually brought back to the coast, where they feed on subtidal molluscs and mussels. Much of our knowledge of these turtles, which are most abundant along the Zululand and Maputaland coasts, is to be found in the work of Dr George Hughes and the former Natal Parks, Game and Fish Preservation Board, who were among the first to study these fascinating reptiles and how they coexist off our shores.

The five species of sea turtle do not appear to compete for food as would be expected but feed at different depths in the ocean to avoid competition. Different trophic levels are utilised, each turtle species occupying a specific niche and the whole system is like a finely adjusted balance, never completely stable,

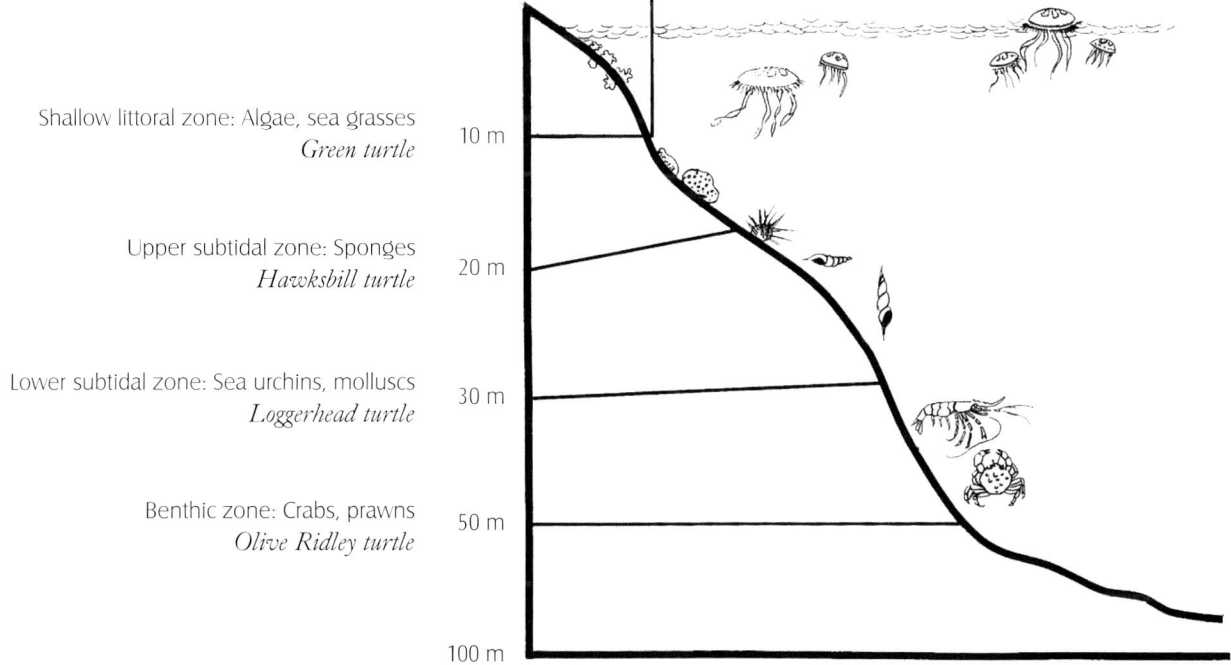

Different trophic niches of South African sea turtles (after Hughes 1982).

always in a state of flux. Green turtles for instance feed on algae and sea grasses in shallow waters, whereas the Hawksbill feeds on sponges in the subtidal zone. The Loggerhead feeds on sea urchins and molluscs in the lower subtidal zone 20 to 30 m below the surface while the Olive Ridley turtle feeds in the benthic zone at depths between 30 and 50 m on crabs and crayfish. The Leatherback turtle and the young of the other species feed on floating or drifting fauna such as jellyfish and bluebottles. Because turtles feed in the ocean, they ingest large quantities of salt which if not voided would cause a serious imbalance in the blood and result in the dehydration of the animal. They are therefore able to excrete salt by means of glands situated in the anterior corners of the eyes, which account for the tears sometimes seen in the eyes of a turtle.

Turtles are also able to spend up to two hours underwater and may even sleep during this time. Prior to diving, the turtle takes several deep breaths and then submerges and finds a place among the rocks where it can lodge itself. Its buoyancy is adjusted internally and it closes its eyes in sleep. A series of involuntary adjustments takes place inside the turtle. First blood is pumped into the nasal membranes causing them to swell and close the nostrils so that water cannot accidentally get into the lungs and drown the animal. Next the brain, about the size of a human thumb in a 250 kg turtle, works out which parts of the animal's anatomy requires a good oxygen supply and then closes sphincters on main arteries, sharply reducing blood flow to the chest muscles, fore and hind flippers and so forth, but keeping a normal flow to the brain. The heart rate gradually drops as the turtle enters into a deep sleep until it is beating at less than 20 per cent of its normal rate. The rate at which waste products of metabolism are produced slows down and since the animal is highly tolerant of toxic byproducts it need not remove them during the two-hour period, which is a reptile underwater record.

THE LEATHERBACK SEA TURTLE FAMILY
DERMOCHELYIDAE

The largest of all sea turtles and certainly the second heaviest living reptile is the Leatherback turtle, *Dermochelys coriacea*, so-called because its carapace is covered not in scales but a smooth leather-like dark coloured skin. It can attain a mass of 750 kg and measure 3 m across the flippers. This amazing animal travels extensively across the world's oceans. One specimen which was tagged in French Guiana, northern South America, was recaptured in Ghana 10 months later having travelled at least 6 000 km.

These turtles also are the only known reptiles to be able to generate internal heat by burning up fat reserves, particularly when undertaking deep dives. They frequently dive to a depth of 60 m but may reach 350 m or more on occasion and are able to stay under water for 37 minutes or longer.

Small numbers nest on the Maputaland coast of northern KwaZulu-Natal, which is one of a few such sites found in the world. A female can only nest up to three times in her lifetime. During the nesting season, each female can lay up to 1 000 eggs in batches of 100 to 120 a time. Each batch is laid at nine- or ten-day intervals, depending on the ocean temperature. The female comes ashore at night during the summer months. As she emerges from the surf she is cautious and on the alert for any possible danger. If satisfied that there is none she will move higher up onto the beach and search for a suitable place to lay her eggs. Having decided on a site she digs a cavity with the fore flippers until she lies in a depression, deep enough so that her carapace is level with the surrounding sand. She then uses her hind flippers to dig an egg cavity, about 45 cm deep and flask-shaped. The eggs are extruded in bursts of one to four and after all have been laid, are gently covered by sand

The Leatherback turtle female coming ashore to nest at Black Rock along the Zululand coast. Photo: W.D. Haacke

until the hole is filled. During this process she continuously feels into the hole with her hind flippers. Finally she adds more sand and then kneads and presses the surface until it is packed hard. Before returning to the sea, she disguises the nest still further by scattering loose sand with the fore flippers. The eggs are highly fertile (ca 75 to 90 per cent) and take up to 70 days to incubate. Hatchlings are only 50 to 60 mm in length and weigh about 40 g. As in the case of all other hatchling turtles, this is the time of greatest danger. They must now dig their way out and make for the sea as fast as possible. Gulls, mongooses, jackals, stray dogs, ghost crabs and many other animals, including ants, prey on the youngsters and it is generally accepted that only 1 per cent of a nest will reach maturity. Hatchling emergence usually takes place at night to avoid predation. Hatchlings grow rapidly and reach sexual maturity at four to five years

A Leatherback turtle coming ashore to nest in the Mabibi area along the Zululand coast. Photo W.D. Haacke

of age and a carapace length of 1 400 mm. On attaining sexual maturity these animals return to nest on the beaches where they hatched.

This is the rarest turtle in the world and it is encouraging to note that while many populations elsewhere are on the decline due to wholesale slaughter for food, their numbers are on the increase on the Maputaland coast, largely due to the protection and care given by the KwaZulu-Natal Parks Board.

THE MODERN SEA TURTLE FAMILY
CHELONIIDAE

The picture is essentially similar with the other turtle species, the most common of which is the Loggerhead turtle, *Caretta caretta*. The adults have no natural predators in the sea with the exception of humans and some of the larger sharks. They are smaller than the leatherback, only rarely reaching a mass of more than 140 kg. They are found worldwide with major nesting sites in Japan, the south-eastern coast of the United States and especially an island off the coast of Oman in the Persian Gulf, which boasts the largest colony with up to 30 000 individuals. They breed along the Maputaland coast where up to 500 females may come ashore to nest.

They range widely and are known to migrate as far as 2 600 km from their nesting sites, one female is known to have completed the journey in 66 days.

Loggerhead females lay on average about 500 eggs per season in batches of 100 to 120 at 15-day intervals. They may come ashore to lay eggs each year or may be absent for as long as eight years and can nest at least five times, but possibly more, during their lifetime. Again, the percentage survival of the young is very low and it is noteworthy that the South African population is one of few, if not the only one, which is protected.

The young tend to travel north to warmer waters in the tropics but some may go south with the Agulhas current, washing up along the southern Cape shores.

Another species belonging to this family is the Green turtle, *Chelonia mydas*, which is perhaps best known as the main ingredient of turtle soup. This turtle is also a visitor to our shores but does not breed here. One of their main breeding sites is on Europa Island in the Mozambique Channel, approximately midway between Maputo and Morombe on the west coast of Madagascar, where between 4 000 and 9 000 individuals breed annually. Nesting takes place throughout the year, with hundreds of females coming ashore each night to lay eggs.

They travel widely, and in fact those nesting on Ascension Island swim up to 2 200 km to the coast of

The Loggerhead turtle, one of the more common turtles off the coast of South Africa. Photo: W.D. Haacke

Brazil where they spend the non-breeding period. Because of the heavy exploitation of the eggs and adults by humans through the years, the distribution of nesting sites has been drastically reduced, and only a small number now remain. However, the species has a high reproductive potential which should allow recovery if some restraint in harvesting is undertaken and populations afforded protection.

Most turtle species are eaten by the people of countries to whose shores they come to breed and are captured in nets while feeding offshore. In Madagascar the turtles were regarded as 'fady' (taboo) by some tribes, but people avoid this custom by placing the head on a stick facing the sea, so that its spirit may continue wandering across the oceans and not place a curse on the fisherman who caught and killed it.

The popularity of turtles as a food item in Madagascar can be seen in a survey undertaken by Dr George Hughes during 1970, which showed that over 23 000 turtles were killed and eaten in that country alone. This is not only peculiar to Madagascar but also occurs along the coast of Africa and worldwide where the adults and eggs are a popular source of food to fishermen.

Other uses include tortoiseshell, which does not originate from tortoises but consists of the lamina covering the bony skeleton of the carapace of the Hawksbill turtle. Turtle leather is another product in demand. In 1969 over 500 000 Olive Ridley turtles, *Lepidochelys olivacea*, were slaughtered in Mexico

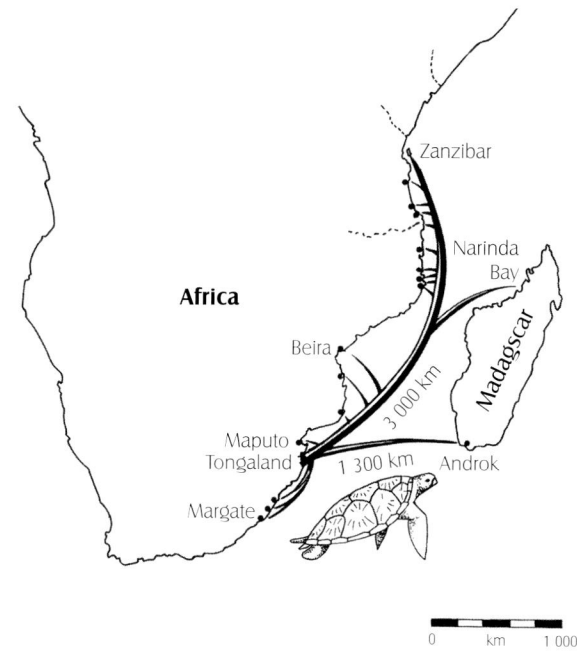

The long-range movements of Loggerhead turtles tagged along the Tongaland coast (after Hughes 1982).

for their hides. The eggs of all sea turtles are eaten by humans, pigs, dogs, mongooses and ants. It is small wonder that many of the species are threatened with extinction. In this respect South Africa is making a positive contribution towards their continued survival, protecting nesting grounds on the Maputaland coast. Research is undertaken by the KwaZulu-Natal Parks Board and the Oceanographic Institute, which together are accumulating data which will help save these ponderous giants of the sea, whose movements and feats of navigation over the oceans of the world are nothing short of miraculous.

A Transvaal dwarf chameleon from the Soutpansberg, initiating prey capture. Photo: Richard Boycott

LIZARDS

ORDER SQUAMATA

SUBORDER SAURIA OR LACERTILIA

Of all the reptiles, lizards are the most versatile and abundant. Lizards can be found almost everywhere. Many species have adapted themselves to man-made conditions, and may be more abundant under these circumstances than in their natural environment. Two such species are the Common dwarf gecko, Lygodactylus capensis, *and the Tropical house gecko,* Hemidactylus mabouia, *which have expanded their ranges and now occur where they were not present in the past. Lizards are very diverse in form and habits. They display an extraordinary range of adaptations to enable them to survive in their varied environments.*

Like most reptiles, lizard species richness decreases away from the equator with relatively few species occurring in the temperate regions. The deserts of Australia are particularly rich in species with as many as 40 species occurring in some areas. In South Africa up to 25 species have been found to coexist. Species richness is a function of habitat diversity but only up to a point, after which environmental parameters become more important.

Thermoregulation

In spite of the fact that lizards do not like cold temperatures a few species have adapted to survive even under subzero temperatures. On the top of the Cape Fold and Drakensberg mountains several lizard species live among the rocks and short grass, which during winter may be covered by snow and ice. One of these, Cottrell's mountain lizard, *Tropidosaura cottrelli*, can be found up to 3 000 m above sea level.

At the opposite extreme some lizards are able to tolerate high temperatures and many have interesting behavioural adaptations to avoid overheating. The critical temperature for most species is about 42 °C, above which they rapidly overheat and die. Lizards are ectotherms and therefore have to increase their body temperature in the morning in order to be able to move fast enough to catch prey. To achieve this they lie in the sun, usually at right angles to the sun's rays so that the greatest body surface area is exposed to the rays. To do this they flatten their bodies either dorso-ventrally, hugging the substrate or stand on extended legs and flatten themselves laterally. Once optimum temperature has been achieved they move about foraging for prey. When the sun reaches its zenith, temperatures in the open will exceed the critical maximum and rise to 50 to 60 °C on the ground. The lizards then move into the shade where they are able to cool down and forage again. This shuttle behaviour allows the lizards to thermoregulate, maintaining their temperature at optimum functionality.

Cottrell's mountain lizard which occurs along the summits of the Drakensberg. Photo: W.D. Haacke.

Anatomy

The skull of the lizard is not very robust and has large openings behind the eye sockets for the jaw muscles. The lower jawbone is hinged to the quadrate bone, which is movable. In fact, the whole of the upper jaw and front part of the skull can move slightly. This movement is termed cranial kinesis and is also present in birds and certain other reptiles. The reason for this is not clear but it could assist in manipulating prey into the mouth.

All lizards have teeth but these differ according to species. Most have teeth on the bones along the edge of the jaw, others also have palatal teeth as are found in amphibians and snakes. The teeth can be sharp and slightly recurved, or cone-shaped, or blunt and peglike. A few lizards such as the agamas have canine-like teeth at the corners of the mouth. There does not appear to be a correlation between the type of teeth and the food eaten. The attachment of the teeth varies from pleurodont to acrodont. It appears that acrodont teeth tend to be permanent whereas pleurodont teeth may be replaced throughout life.

Smell

Lizard tongues are highly variable. Monitor lizards have split snake-like tongues, whereas the skinks have short blunt tongues. In between there is a range of variations; geckos for instance have long flexible tongues so that they are able to clean their eyes by licking them, for they do not have eyelids. In all lizards the tongue is covered in papillae and has functions of taste, touch and smell.

The tip of the tongue brings scent molecules to the paired Jacobson's organs in the roof of the mouth. These are well developed in lizards and snakes as well as in other reptiles and are present in most mammals, although not as sophisticated. They are situated on the roof of the mouth and are in contact with it by a narrow duct, which opens near the front of the palate, slightly off-centre from the midline. These organs are lined with sensory cells similar to those found in the nose of mammals and are activated by a branch of the olfactory nerve. Minute particles are picked up by the tip of the tongue and are brought into contact with the sensory surfaces of these organs, where they activate the sensory cells lining their cavities. These organs play an important role in locating food and during the breeding season when seeking a mate as well as in a variety of other ways. In certain lizards such as chameleons these organs are not as well developed because these animals hunt by sight and not by smell.

Caudal autotomy

The ability of lizards to lose their tails is well known. Geckos have soft tails and tend to shed them more easily than other lizards. This self-mutilation or autotomy occurs when the muscles of the tail contract. The ability to shed the tail depends on a series of weak fracture planes. These planes pass through the individual vertebrae and not between them. As the break occurs, the surrounding muscles separate neatly at the fibrous partition between two of the segmented muscle nodes, with very little loss of blood. Regrowth of the vertebrae does not take place but instead is replaced by a cartilaginous rod, which supports the new tail. Sometimes the tail only breaks on one side, in which case it often happens that a second tail grows out from the break line. A regenerated tail can be easily recognised as it differs in scalation and colour from the original. Caudal autotomy is not only a defensive measure against predation but may also occur during intraspecific fighting, that is between two individuals of the same species. A survey of regenerated tails of lizards in the northern provinces ranged from 40 to 78 per cent for geckos and skinks, 7 to 16 per cent for lacertids, 14 to 40 per cent for gerrhosaurids and between 24 and 62 per cent for cordylids, indicating the degree to which these animals make use of this ability.

Food

Lizards are for the most part opportunistic feeders. Some are more specific and feed mostly on termites while others prefer ants. In the main, lizards feed on a variety of prey according to the size of the species. Most feed on invertebrates, some are primarily insectivorous, a few may include small vertebrates, some even including other lizards in their diet. Cannibalism occurs when the adults of some species may include the hatchlings of other species in their diet, or feed on other small lizards. Most species feed on what is available and palatable. This is understandable, since being dependent on a narrow range of prey may make them very vulnerable. Should such prey decline or disappear then the predator is likely also to be affected and may even be threatened with extinction. Therefore versatility is necessary and the fact that lizards are numerous and widespread indicates their success.

They have adopted two main foraging strategies in locating prey: one consists of actively seeking or hunting for food while the other is passive, waiting for potential prey to pass by specific vantage points. Lizards feed mostly by day with the exception of most geckos and a few lizards which are nocturnal hunters. Active foragers tend to be swift-running lizards whereas both the 'sit-and-wait' and nocturnal foragers tend to be slow moving, although able to produce a burst of speed when the prey comes within striking range.

Lizards are the prey of a large range of predators, including other lizards, snakes, birds, mammals and even invertebrates. How do lizards escape predation? Many species such as geckos are very cryptic, moving like shadows on rocks and trees. If approached they run into the shade where their colouring blends in with the background, or they move crab-like around the bole of a tree or a branch, thereby always keeping the tree between themselves and the threat. When disturbed or annoyed some lizards open the

The regenerated tail of this Variable skink, Trachylepis varia*, has not broken off completely with the result that another has developed along the fracture forming a forked tail.*

Predation by a Lycosid spider on a Cape dwarf gecko.

mouth wide, displaying the yellow or black interior which is designed to frighten off would-be predators by making the lizard appear larger and more fearsome. At the same time the lizard will adopt attacking behaviour, which includes moving jerkily – the suddenness of these movements is intended to confuse the predator causing it momentarily to back off, giving the lizard a chance to escape.

BEHAVIOUR

Many lizards are territorial, in other words they actively defend an area against others of the same species. This type of behaviour is usually associated with males but is also applicable to females, who may assist the male on occasions. Some lizards, especially the males, are brightly coloured. These colours are usually blue, green, black, yellow, red and white, which indicates that animals of the species are not colour blind. The male usually displays its colours from a select vantage point within or along the boundary of its territory. He usually adopts a specific pose to display his colours to maximum effect. Only males with adult mature colours do this, which conveys the message that a dominant male occupies this territory and serves as a deterrent to others of the same species. This behaviour appears designed to avoid conflict. However, should another adult male ignore the warning signs and enter, he is then issuing a challenge to the resident male, and should the former not move out of the area a fight will ensue. Such fights can be quite vicious and the victor may chase the defeated rival for some distance. The desire to terminate the fight is usually shown by the flight of the defeated animal.

Such challenges may take the form of a lateral display whereby the animals present the bright colours of the sides to each other and the two males circle each other until one rushes in to bite its opponent. They grapple, rolling, twisting and biting while each one tries to turn the other over until one of them decides he has had enough and flees.

Within a dominant male's territory there may be several subordinate males, usually with colours less bright or different colour combinations than that of the adult. They remain in his area until fully mature and able to challenge the territorial male. So in some lizard species there is a hierarchical system in which a dominant male has several females, immature males and juveniles within his territory. The animals are subservient to him and relinquish choice basking sites and even food should he be around.

In some species the colour is much brighter during the breeding season and the males flash these colours to attract females and at the same time deter other males. Apart from colour, males of some species are considerably larger that the females. Typical examples include the Tree agama, *Acanthocercus atricollis*, and crag lizards. Males of the former have bright blue heads which they bob up and down, or else achieve the same effect by doing press-ups, bending and straightening the elbow.

The defensive posture of the Giant ground gecko with arched tail and open mouth, with which it can inflict a painful bite. Photo: Richard Boycott.

Reproduction

Courtship in lizards is usually not an elaborate affair. The male will approach the female with jerky movements, constantly flicking his tongue out and touching the female. She moves about unconcerned and often the male ends up chasing after her. He bites her in the neck and twists his body under hers so that the cloacas are in contact. Males have two hemipenes but only one is used at a time. In many species these are adorned with spines and calyces or growths, which assist in keeping the hemipenis in place during mating. These growths usually differ from species to species and can be a valuable taxonomic tool. The animals remain locked in this position for varying lengths of time. Once fertilisation has taken place the male departs and has no further interest in the female. He may mate with several females at this time. Little grooming takes place afterwards although the cloacal region may be licked.

Lizards may be oviparous or viviparous, although an in-between condition called ovoviviparity also occurs in some species whereby the eggs are retained within the body of the female for varying lengths of time during which the embryo develops, nourished by the egg yolk. The egg with the almost fully developed embryo enclosed by the membrane, is laid in a suitable site. On completing development, an egg tooth enables the embryo to slit the membrane and it is born fully developed. This is similar to viviparity where the embryo is fed via a placenta, except here the embryo is not directly linked to the female, but only incubated inside the body.

In oviparous species, after fertilisation the eggs develop and prior to laying are coated with a flexible membrane, or a rigid membrane in some species. The female then searches for a suitable nesting site. Having found one she begins to dig, scratching out a hole at an angle. This is accomplished by the forefeet, which are armed with claws. The hole varies in depth depending on the species and size of the lizard. However, the hole is normally deep enough to allow the lizard to lie in it full length while she lays her eggs. During the digging process she frequently enters the hole and turns around inspecting it as if assessing its specifications. When the hole is completed she lies inside and proceeds to lay her eggs at a rate of one per minute or at longer intervals. When all have been laid she emerges and begins to scratch the excavated soil into the hole and fills it up. Every now and then she pauses to tamp the soil firmly with her snout until the soil is packed hard. She then scratches about and attempts to disguise the site by scattering loose soil. Sometimes she will leave the nest for a while before returning to make additional finishing touches. This can happen several times. The incubation of the eggs is variable according to temperature and time of year of laying and may take up to 10 months in some species such as monitor lizards and Common flap-necked chameleon, *Chamaeleo*

The Drakensberg crag lizard, showing colour differences between male (left) and female (right).

Eggs and hatchling of the Yellow-throated plated lizard.

dilepis. Low temperatures retard the growth of the embryo and high temperatures accelerate growth. Temperatures of 25 to 30 °C need to be maintained during incubation.

Once the young hatch they dig their way through the soil, which has been moistened by egg residues, to the surface where they quickly disperse. The eggs and young prior to hatching are very susceptible to fungal attack, ants and other predators such as mongooses. It has been observed in West Africa that in two species of lizards only 10 to 20 per cent of the eggs develop to the adult stage. To offset this, some lizards lay large single clutches while others lay up to three small clutches in a season, while other species breed throughout the year.

Viviparous lizards do not require special nest preparations and the female gives birth at any convenient and protected site. No further interest is paid to the young after birth, as they are able to fend for themselves.

LONGEVITY

Longevity of South African lizards is generally difficult to determine in the wild and little information is available. Some species only live for a year and others for longer. Tagged Veld monitors, *Varanus albigularis*, have been seen two years after being first tagged but their age at that time was unknown. In captivity some lizards such as the Water monitor, *Varanus niloticus*, plated lizards, *Gerrhosaurus flavigularis*, *G. validus* and *G. major* and flat lizards, *Platysaurus guttatus*, have reached in excess of 10 years of age, some living up to 15 or 16 years, and many other species have been recorded to live between five and 10 years and a larger number below this.

The Agama Family
Agamidae

This family of lizards is known to most people and includes some very interesting lizards that live on trees, rocks or are terrestrial. They are, on the one hand, the most conspicuous and, on the other hand, the most inconspicuous lizards found in the veld, depending on the time of year and circumstances. Two genera and six species with one subspecies are known to occur in South Africa, most of which are terrestrial. Among the most familiar to people living in the northern parts of South Africa is the Tree agama, *Acanthocercus atricollis*, or Bloukopkoggelmander, which is typically seen on the boles of large trees in the bushveld. They are relatively large lizards, exhibiting sexual dimorphism with males reaching 167 mm snout-vent length whereas females only reach 135 mm. This difference also extends to colouring, with breeding males having blue-green, blue and turquoise coloured heads whereas females and non-breeding males are cryptically coloured tending to be olive brown to blackish above. If a male challenges another male or tries to impress a female, his head will be bluest, while the body displays different colour combinations of green, white, brick red, blue and black. However, should his life be threatened then his colouration becomes as cryptic as that of the female who is normally nearby, clinging motionless to the bole of the same tree. These lizards occasionally fight during the breeding season and tails may become scarred as a result. Such injuries are the result of canine-like teeth found at the corners of the mouth, which they use in defence or to deliver a painful bite. However, despite numerous tales to the contrary a bite from these lizards is not poisonous. When threatened they gape widely showing off the bright inside of the mouth as a deterrent, and may attack if this warning is ignored. These lizards are unable to regenerate their tails following amputation by predators and many individuals have been encountered with stumpy tails as a result.

These lizards live primarily on large trees, and are usually seen in an upright position on the bole of the tree, frequently nodding the head or bobbing up and down at the elbows. They rarely descend to the ground but take refuge under loose bark or in holes or crevices in the tree trunk. On occasion they run across open space to another tree or when hunting prey that passes a short distance from the base of the tree.

They forage from a vantage point, usually hanging close to the ground on a tree trunk watching for prey activity below whereupon they dash down and across the ground to snap up the prey. They are opportunistic feeders and will consume grasshoppers and beetles but ants and especially termites and their alates are preferred when available. One individual appeared to feed extensively on bees, with nine stings found imbedded in its mouth. They will also take caterpillars and perhaps small vertebrates.

They breed during the summer months and after mating the female lays five to 14 eggs measuring 10 to 21 mm × 22 to 28 mm, in a hole dug by the female. The eggs hatch after about 90 days and the hatchlings, which measure about 70 to 80 mm, emerge and disperse. Growth during the first three months is rapid and they may triple their length but this slows down as

Male Tree agama displaying his brightly coloured head and body to maximum effect.

A Common ground agama perched on the top of a shrub surveying his territory.

A territorial male Ground agama and challenger attempting to intimidate each other by circling with maximum lateral displays.

A resident territorial male displaying to a challenger.

the animals get older. They are sexually mature at two years of age.

Common to the Northern Cape and further north, the Ground agama, *A. aculeata* is widespread in South Africa extending from the Kalahari in the west to KwaZulu-Natal in the east. Two subspecies are currently recognised: Ground agama, *A. a. aculeata*, and Distant's ground agama, *A. a. distanti*, the latter smaller than the former and mostly found in grassland and bushveld of the more northern provinces with the typical form present in the Kalahari, much of the Northern Cape, most of Botswana and most of Namibia. The Ground agama is terrestrial and lives in holes under bushes dug by rodents or by the lizards themselves. Like the previous species, the males develop a blue head during the breeding season but this does not reach the same brilliance. Nevertheless, at this time the lizards may often be seen in the heat of the sun, sitting on top of fence poles or bushes showing off their colours to impress males and females alike. At this time, aggressive displays and fighting may take place. The males usually approach obliquely circling each other so that the side of the body and distended throat are presented to the opponent in order to intimidate him. If this fails then the males will fight until one decides he has had enough and flees.

Such vantage points may be used for basking and thermoregulation as well. These lizards are very cryptically coloured and when disturbed may run off rapidly for a short distance, usually into some brush or other cover and then lie down hugging the ground to escape detection. Often when running at speed the tails are arched upwards, especially in Distant's ground agama. Their prime enemies appear to be the smaller raptors, mongooses and meerkats.

These lizards feed mostly on ants as well as on termites and their alates and will spend lengthy periods feeding on such small prey, but they also eat other invertebrates.

The Ground agama is oviparous and the female lays seven to 18 eggs, measuring 12 to 16 mm × 9 to 11 mm, during midsummer. Incubation may be rapid and at 30 °C ranges from 45 to 50 days. Hatchlings measure between 56 and 65 mm in total length.

The Rock agama, *Agama atra*, is another widespread agama species in South Africa occurring from north of the Soutpansberg to Cape Town. It occurs from sea level to over 3 000 m in Lesotho, enduring extremes of temperature. Two subspecies are recognised: the widespread nominate race *A. a. atra* and *A. a. knobeli* the latter restricted to Namaqualand and southern Namibia. Both are a purely rupicolous or rock-living species occurring in small groups where suitable refuges in the form of crevices and holes under and between rocks are found.

These lizards are usually seen basking on top of boulders or bobbing up and down signalling to other members of the group. They are particularly agile on the rocks and leap across chasms from one side to the other. Breeding males have a blue head which, like that of the other species, is flaunted from the top of a boulder or some other vantage point as a signal to male and female alike, although with different intentions. However, if danger threatens the colours fade and the animal becomes cryptic, hugging the rock and becoming almost invisible. They take refuge in vertical crevices if pursued or at nightfall and wedge themselves in, resisting attempts to dislodge them.

Males tend to be larger than the females and other subordinates, reaching 122 mm in snout-vent length whereas females may reach 99 mm. Both males and females maintain territories, which in the former is larger and has been recorded as being up to 90 m^2 in extent, usually incorporating the territories of several females. Both sexes form hierarchies with a dominant male and female and many subordinates. These lizards occur in considerable densities, up to 165 individuals per hectare, in optimum habitat.

Like the previous species they also mostly consume ants and termites but will also feed on a variety of other invertebrates such as grasshoppers, beetles, crickets, millipedes and spiders.

The breeding season is during summer and two clutches of eggs may be laid, one during November and December with a possible second one during January to February. Clutches range from five to 18 eggs, with eggs measuring

A male Rock agama in cryptic colouration.

15 to 18 mm × 10 to 12 mm. These are laid in a hole in the ground and hatch after an incubation period of two to three months. Hatchlings measure 59 to 68 mm in total length.

The agamids are widespread all over the world and some quite bizarre forms are found, especially in Australia where 'Thorny devils', Frilled and Bearded lizards and Water dragons are a few of the more spectacular species. The only flying lizards in the world belong to this family.

This is a unique lizard, which can take long gliding leaps through the air by means of wing-like extensions of the skin along the flanks. These 'wings' are supported by elongated ribs and can be spread or folded at will, but cannot be flapped. In some species they are brightly coloured and displayed in courtship and in flight. These lizards are relatively small reaching about 30 cm in length. The Flying dragons, as they are called, live in the forested areas of south-eastern Asia.

THE GECKO FAMILY
GEKKONIDAE

Most people are familiar with the lizards of the family Gekkonidae or geckos. The name gecko is derived from the call of one of these lizards from south-east Asia. There are numerous species which vary greatly in size, from the relatively large Turner's thick-toed gecko, *Pachydactylus turneri*, and Wahlberg's velvety gecko, *Homopholis wahlbergi*, to the small dwarf and leaf-toed geckos. Geckos are distributed throughout the warmer regions of the world and number about 760 species of which at present about 59 species and several subspecies occur in South Africa, many of these endemic. They occupy all habitats from tidal marshes to above 2 000 m on the highveld of Mpumalanga.

Geckos are characterised by having a brille covering the eye instead of an eyelid and most species exhibit expanded tips to the digits, which are covered by rows of scales or lamellae. These scales are furnished with minute hairlike projections which fit into irregularities in the surfaces to which the animals cling. If a sheet of glass is magnified a hundred times these irregularities, which the gecko uses to climb up apparently smooth surfaces, become visible. Some geckos do not have these expanded tips, the digits simply terminating in claws. Dwarf geckos of the genus *Lygodactylus* have lamellae under the tail tip which assist these arboreal and rupicolous geckos to cling to vertical surfaces. A very unique adaptation is the webbed feet of the Namib web-footed gecko, *Palmatogecko rangei*, and the Kaoko web-footed gecko, *Kaokogecko vanzijlii*. The former just enters South Africa along dunes in the Richtersveld whereas the latter is restricted to the northern Namib. These geckos possess webbing between the fingers and toes for easy travel across the loose sands of the desert permitting them to climb up steep dunes. They also enable these geckos to dig burrows in loose sand.

There are three species of Barking gecko, namely Koch's barking gecko, *Ptenopus kochi*, Carp's barking gecko, *P. carpi*, and the Common barking gecko, *P. garrulus*, the only one to occur within our borders. These are very quaint lizards, which have a wide distribution in Botswana and Namibia but are restricted in South Africa to the Limpopo and Northern Cape provinces. They live in colonies and are a very vocal species. These geckos do not bark but actually utter nasal clicking calls,

Wahlberg's velvety gecko, a rock-living species also often found around houses and outbuildings. Note the dark interior of the mouth and regenerated tail.

A Common barking gecko at the entrance of its burrow.

which on a quiet night may become deafening when one has established a camp close to a large colony. They live singly in self-excavated burrows, which are about 30 cm in depth usually with a twist along the way and often with one or more false passages, which terminate just below the surface. The males have a bright yellow throat and when calling lie at the entrance of the burrow, which may help to resonate the call further. They are most active at dusk but in some areas may call throughout the night. In the Vivo area calls are restricted to the summer months whereas in the Kalahari they call almost throughout the year and are an endearing feature of Kalahari nights. If one sits quietly with a torch near a burrow the animal may come out and feed on insects attracted to the light.

Most geckos are active at night but the dwarf geckos form a group of diurnal species, which are primarily active by day although this may extend into dusk. There are eight species and several subspecies of these geckos in South Africa, five of them endemic, mostly concentrated in the northern provinces.

The most common and widespread of these is the Common dwarf gecko, *Lygodactylus capensis*, which although being primarily an arboreal or tree-living species also commonly occurs on the walls of houses and on rocky outcrops. However, they are most abundant on trees and dead wood lying about on the ground, especially those with holes and rough or flaking bark under which the animals can take shelter at night or when rain threatens. They tend to forage from perches close to the base of the tree but also cross open spaces on the ground from one tree to another.

These geckos are very cryptic in appearance and behaviour and blend in very well with the bark of trees and dead wood on which they live. In addition they appear able to lighten or darken their colour to blend in better with the surroundings. Their habit of sitting and waiting for prey makes them difficult to see but they become more visible when they change position. If approached they either flee up the bole of the tree or they move crabwise around the bole of the tree keeping this between the attacker and itself. They are communal animals living in colonies where the dominant male usually has the best basking, escape and foraging sites. Young individuals appear to be relegated to smaller and less secure habitats within the territory of a dominant male. They may often be seen lashing the tail sideways, usually a signal to other conspecifics (individuals of the same kind) to not approach closer or risk combat with the dominant individual in the group.

The Common dwarf gecko is an inveterate hitch-hiker and by taking advantage of humans and their desire to travel has established populations far from its original home, riding in vehicles or other goods and descending where the vehicle stops for any length of time. In recent times populations have been recorded from the Addo National Park, many urban centres such as Bloemfontein and Port Elizabeth and now even in George in the Western Cape. It seems that this species has become a commensal, benefiting from human activities and in time may establish itself throughout the country, perhaps to the detriment of some endemic species.

The largest of these geckos is Methuen's dwarf gecko, *L. methueni*, which is also at least partly arboreal. It is an attractive velvety olive green in colour with many markings and is only known to date from the Woodbush Forest, an area of roughly 10 000 ha in the vicinity of Haenertsburg in Limpopo Province. The species appears to lay its eggs in crevices of boulders and the hatchlings have orange-red tails. The South African *Red Data Book on Reptiles and Amphibians* lists this species as vulnerable, as much of its former habitat was destroyed during commercial afforestation of the region.

All of the remaining species and subspecies with the exception of Bradfield's dwarf gecko, *L. bradfieldi*, which is also arboreal, tend to be rupicolous or rock living but will ascend trees on rocky outcrops. These include the Ocellated dwarf gecko, *L. ocellatus*, the Black-spotted dwarf gecko, *L. nigropunctatus*, Stevenson's dwarf gecko, *L. stevensoni*, Waterberg dwarf gecko, *L. waterbergensis*, and the Granite dwarf gecko, *L. graniticolus*. Some of these, such as the Waterberg and Granite dwarf geckos, have very restricted ranges in Limpopo Province, occupying the upper slopes of the Waterberg and the granite hills around Makopane

The Common dwarf gecko is widespread in the northern and eastern parts of South Africa and is currently extending its range.

An unusually coloured individual from the Drakensberg escarpment in Limpopo Province.

TOP: *The Woodbush dwarf gecko, a rare species endemic to Limpopo Province.*

CENTRE: *The endemic Waterberg dwarf gecko is restricted to the Waterberg massif in Limpopo Province.*

BOTTOM: *The Tropical house gecko is an inveterate hitch-hiker.*

(formerly Potgietersrust) respectively. However, both these lizards occur in protected areas and are currently not endangered.

The Ocellated dwarf gecko is restricted to the highveld and escarpment regions of Mpumalanga with small outliers in Gauteng and another subspecies *L. o. soutpansbergensis* to the frontal range of the Soutpansberg.

These geckos are perhaps even more cryptic than the Common dwarf gecko as they tend to be smaller and are more flattened dorso-ventrally – an adaptation to a rupicolous way of life. They are so well camouflaged that they blend in with the rocks on which they live. They are only observed when they move like a shadow around the side of the rock or boulder and disappear into more shadowy depths. Like all dwarf geckos they lay two eggs, usually glued together at the widest point, at the base of a boulder adjacent to a grass tussock.

Another species that is prone to hitch-hiking is the Tropical house gecko, *Hemidactylus mabouia*, which formerly only occurred in the bushveld and lowveld. It is now common around Pretoria and has also been seen in the Free State and many Eastern Cape towns. It is an almost cosmopolitan species that also occurs along the coast of South and Central America and on many islands in the Indian Ocean. This lizard has adapted well to living with humans by establishing residence in man-made structures, especially houses, where they forage around outside lights and windows feeding on insects attracted to the light. This ability to adapt has enabled this gecko to live in parts of the country where under natural conditions they could not survive during the cold winter months. They have an ability to lighten or darken their skin colour according to the colour of the background, becoming translucent at night. Another interesting trait of this gecko is that it is quite vocal, the voice a click, which may frequently be heard at night and even on overcast days from a distance of several metres.

Probably the largest number of species of any of our reptile genera belong to the genus *Pachydactylus* or thick-toed geckos of which 19 species occur almost throughout South Africa with the exception of the KwaZulu-Natal Drakensberg and parts of the

TOP LEFT: Van Son's thick-toed gecko, a terrestrial species from the lowveld of Mpumalanga.
TOP RIGHT: A hatchling and eggs of Vanson's thick-toed gecko.
BOTTOM RIGHT: The Tiger thick-toed gecko, which mostly occurs north of the Soutpansberg into southern Zimbabwe.
BOTTOM LEFT: The Spotted thick-toed gecko is a peculiar short tailed species mostly occurring along the eastern coastal plain.

highveld of the Free State, Gauteng and Mpumalanga. The greatest number of species occur in the arid zones of South Africa, mostly in the Western and Northern Cape provinces of which 10 are endemic, meaning they do not occur elsewhere in the world. They are either terrestrial or rock living, the former taking refuge in burrows, mostly those of other animals, in termite mounds, hollow trees, under stones, in rotting plant material and in crevices between rocks. One species, the Spotted thick-toed gecko, *P. maculata*, has been recorded taking refuge in the empty shells of the Giant land snail in the Eastern Cape but elsewhere finds shelter under rocks, in rotting logs, aloes and holes in the ground. Some species can be quite vocal, such as the Transvaal thick-toed gecko, *P. affinis*, which often enters houses and can be heard squeaking at night.

Some of our largest geckos belong to this genus, notably Bibron's thick-toed gecko, *P. bibronii*, and Turner's thick-toed gecko, *P. turneri*, which attain snout-vent lengths of 100 mm and 95 mm respectively. A bite from one of these lizards can take the skin off a finger should one handle it carelessly. The former is endemic to the Northern Cape Province while the latter has a larger distribution in the northern provinces extending into Central Africa.

Almost as big, the Velvety gecko, *Homopholis wahlbergii*, is widespread in the Limpopo, Mpumalanga and KwaZulu-Natal provinces. It tends to inhabit crevices between boulders on large rocky outcrops and may be seen basking in the late afternoon sun on the rock faces around its retreat. This gecko also frequents human habitation living on the walls of houses usually under the roof where it forages for invertebrates attracted to lights and windows. They squeak when handled and twist and bite vigorously. The inside of the mouth is a blackish grey. One of the more intriguing aspects of this gecko is that its skin may tear as an escape measure. A much rarer species is Muller's velvety gecko, *H. mulleri*, which has only been recorded inhabiting holes in large trees especially Marula, *Sclerocarya birrea*. It is a much smaller species and is characterised by having a white upper lip.

Probably the most impressive of our geckos is the Giant ground gecko, *Chondrodactylus angulifer*, which has a large head and body to match. A nocturnal lizard, it rests by day in burrows in the ground, as much as a metre in length and usually dug by itself, or sometimes

TOP: *The Giant ground gecko of the Northern Cape.*
BOTTOM: *A lateral view of the head of the Giant ground gecko. Note the narrow slit-like pupil with pinholes for vision in bright light.*

in burrows of other animals such as scorpions, dung beetles and rodents. A solitary species, it moves about at night usually walking slowly but can run quite fast should the need arise.

It feeds on a variety of prey, mostly invertebrates such as Harvester termites, *Hodotermes mossambicus*, tenebrionid beetles, grasshoppers, moths, solifugids, spiders and scorpions but may also include other geckos in its diet. This is a very charismatic species both in appearance and in demeanour. If confronted it stands on stilted legs and arches the tail over the back, presenting as imposing an appearance as possible. At the same time it inflates the body, utters a continuous wheezing hiss, and sways from side to side. If molested it may lunge at the attacker and bite, which may draw blood. The doyen of South African herpetologists Dr V.F.M. FitzSimons recorded in his book *The Lizards of South Africa* that the species was considered poisonous by the human inhabitants of the areas where they occurred, despite it being quite harmless.

A unique gecko is Peringuey's coastal leaf-toed gecko, *Cryptactites peringueyi*, which was until recently only known from specimens collected at Chelsea Point, Port Elizabeth, and Little Namaqualand, two widely divergent localities and described in 1910. Dr V.F.M. FitzSimons, in his Monograph *The Lizards of South Africa*, regarded this species as perhaps not belonging to the South African fauna but rather to the American fauna. It was for a long time considered as one of South Africa's herpetological mysteries. Frequent searches for this tiny lizard, with a snout-vent length of less than 30 mm, drew a blank. It was only rediscovered by accident in 1992 inhabiting a salt marsh along the estuary of the Kromme River in the Eastern Cape Province, living under debris in areas inundated at springtide. It is now known to inhabit the coastal zone and estuaries from Chelsea Point to the Kromme River estuary.

Widespread along the mountains from the Western Cape Province to the Limpopo River, twelve species of unique rock-living geckos of the genus *Afroedura* or flat geckos occur in South Africa. Their distribution follows that of the great escarpment, extending further inland in the northern provinces where a radiation of species has taken place, many of which are as yet undescribed. Most species have very restricted distributions, some virtually occurring along a single range of hills. Most species have developed a flattened body to enable them to occupy narrow crevices between and under rocks, two species also being found under the dry leaves of aloes or living on the walls of huts and houses. Like many other geckos they have the ability to lighten or darken their body colour by the expanding and contracting the black pigment cells called melanophores, and may at times appear translucent especially against a light, as is recorded for the Tropical house gecko. Individuals of some species live solitarily in crevices while others may be communal with many individuals occurring under a rock flake.

The Woodbush flat gecko, Aroedura multiporis, *is endemic to a small area in Limpopo Province.*

The Namaqua day gecko, a South African enigma.
Photo: W.D. Haacke.

A peculiar herpetological mystery of the South African herpetofauna is the occurrence of a species of day gecko, genus *Phelsuma* in Little Namaqualand and the Richtersveld where it inhabits rocky outcrops. The Namaqua day gecko, *Phelsuma ocellata*, is the only member of its genus, which encompasses some 30 species, occurring on the west coast of Africa and the only species to be found in South Africa. Most of the known species occur on the island of Madagascar and other Indian Ocean islands with a few species reaching the east coast of Africa where they are restricted to a moister climate. Most species are arboreal while the Namaqua day gecko is rock living. How and under what conditions this distribution took place can only be speculated upon, as Madagascar split off from the African continent some 60 million years ago. Many reptiles have been known to have rafted to islands on debris washed out to sea during floods and carried away by ocean currents to remote destinations but it appears unlikely that the ancestor of this species rafted on to the west coast as the Agulhas and Benguella currents diverge on either side of the peninsula and it is difficult to see how this lizard could have been transported up the west coast. The alternative theory would indicate an existence of the lizard when Madagascar and Africa were still part of Gondwanaland. As no other species belonging to this genus occurs in South Africa it is difficult to postulate how this species came to be on the west coast and has survived there, while any other potential populations to the east have died out.

THE CHAMELEON FAMILY
CHAMELEONIDAE

The chameleons are a unique family of lizards, highly specialised for an arboreal existence. Almost all chameleons (about 130 species) occur in Africa and Madagascar. Madagascar in particular is rich in numbers of species. A few species occur in Spain, certain Mediterranean islands, North Africa, Arabia and India.

At least 18 species occur in South Africa, mostly dwarf chameleons of the genus *Bradypodion* but including two species of *Chamaeleo*. The dwarf chameleons are endemic to South Africa, many with very localised distributions. Many species are as yet undescribed, mostly along the Drakensberg escarpment of Mpumalanga but also in KwaZulu-Natal and the Free State.

South African chameleons are mostly small to medium sized lizards ranging from 8 to 30 cm in total length whereas some from East Africa and Madagascar reach 60 cm and are capable of feeding on small birds and mammals. They have many features suited to a tree-living existence. The feet of chameleons are unique, the fingers and toes of each limb being webbed and arranged in opposing groups with three toes on the outside and two inside enabling the lizard to clasp round branches, thereby being able to sustain a firm grip even during very windy

A lateral view of the head and eye of the Common flap-necked chameleon, a remarkable adaptation for an arboreal existence.

TOP: *The typical colour of the Common flap-necked chameleon.*
BOTTOM: *The remarkable tongue of the Common flap-necked chameleon is adapted for prey capture in trees and shrubs.*

conditions. The animals are able to sleep in the trees clinging to branches very much like birds do. In addition their tails are mostly prehensile assisting the animal constantly when walking along thin branches, mostly as a balancing organ but also to help anchor the animal during windy conditions. They are very interesting to observe because they move very slowly, often rocking backwards and forwards during this progression. If they become aware of a potential threat, they will sway sideways imitating a leaf shaking in the wind.

When foraging, the lizard progresses just as slowly and the conical protruding eyes are swivelled about independently in any direction for maximum field of vision. Once an eye has located possible prey, the head is turned to face the object and the other eye is brought into focus on the prey. This enables binocular vision and depth of field is achieved. The animal approaches gradually until the prey is within reach of the long distensible tongue, which in some species may be as long as the body and tail. The tongue is brought forward to the front of the mouth and poised there for a moment before being shot out to catch the insect on its sticky expanded tip. This performance is brought about by a specialised musculature of the tongue and head.

A pair of Namaqua chameleons mating.

Many chameleon species have various horns and knobs on the snout while the top of the head is raised to form a bony casque. Such projections are particularly well developed in males and are used in fighting. With the exception of bony casques, South African species do not exhibit such ornamentation. Most species tend to be solitary although the dwarf chameleons may occur in local concentrations but avoid each other. When two individuals meet they tend to avoid head-on confrontations, instead turning broadside to each other and inflating the body and throat for maximum effect. Under these conditions the lizards often adopt bright colours, especially male dwarf chameleons, and may open the mouth to display the interior in a show of defiance.

Chameleons are best known for their ability to change colour to suit the conditions under which they find themselves. This is due to special pigment cells called melanophores, which contain a granular black pigment, and to chromatophores, which contain red and yellow pigments, as well as iridescent cells, which contain minute guanine crystals.

The melanophores are controlled by nerves and are stimulated by stimuli received through the eyes and which convey to the animal the colour of the background. The expansion of the melanophores serves to darken the skin, at the same time obscuring the bright pigmentation. Depending on the overlay of the different pigment cells various colours are adopted. Therefore in green vegetation the lizard is green and when on dry leaves or bark it changes to brown or even black. The latter colour is adopted by the Common flap-necked chameleon, *Chamaeleo dilepis*, when basking on cold winter mornings and serves to absorb the sun's rays so that the lizard heats up rapidly to a stage when it can become mobile. While basking the lizard will flatten its body laterally and orientate itself broadside to the sun, exposing the largest surface area.

Different colour combinations are also possible so that the animal blends as naturally as possible with its surroundings. This is an important survival strategy for such slow-moving creatures. At night these lizards adopt a yellowish green colour while clinging to the branches of a tree or bush. As mentioned earlier, when chameleons are irritated or when fighting or threatened they will often adopt brilliant displays. The dwarf chameleons adopt some very spectacular colour combinations under these circumstances, of which one of the brightest is the Transvaal dwarf chameleon, *Bradypodion transvaalense*, from Haenertsburg, Woodbush and the Soutpansberg. The Common flap-necked chameleon, however, turns an unpleasant blackish colour, simultaneously inflating its body and showing the bright yellow inside of the mouth, while it hisses loudly in an effort to intimidate its opponent. They bite freely and a painful nip can be experienced even to the extent of losing some skin if bitten by a large specimen. They are, however, completely harmless and should be commonplace in every garden but are frequently killed while crossing roads and are predated on by domestic animals such as cats and dogs and are also killed by pesticides used to control garden pests. Under natural conditions the Tree snake or Boomslang, *Dispholidus typus*, is an important predator. Veld fires must impact considerably on chameleons, and dwarf chameleons in particular, as many species inhabit grasslands and are often found clinging to the tops of grass culms.

Many chameleon species lay eggs, including the Common flap-necked chameleon, usually during summer, in holes dug in the ground by the female to a depth of 8 to 10 cm. Here they incubate for up to 90 days. In contrast the dwarf chameleons are viviparous producing up to 17 young after a gestation of about a year.

Many species from different genera have subsequently

become terrestrial. Both Madagascar and South Africa each have a unique ground-living species, both of which inhabit dry areas. The Namaqua chameleon, *Chamaeleo namaquensis*, inhabits very arid areas including true desert and is distributed from the Karoo further north to Namaqualand, the Namib Desert and into southern Angola. It is a unique species that feeds mostly on invertebrates such as locusts and beetles but may take other lizards and even small snakes, including Peringuey's adder, *Bitis peringueyi*. When food such as beetles is plentiful the lizard may consume large numbers and conversely when prey availability is low, it is able to exist for lengthy periods without eating. During the heat of the day the lizard may climb up into a bush or on a rock in order to cool off, a strategy adopted by several other lizard species. A solitary species, the Namaqua chameleon is territorial, both females and males defending their territory against intruders. Males and females come together only when mating. The male, being smaller than the female, may cling to her back during mating. Females may lay up to three clutches of eggs a year.

THE SKINK FAMILY
SCINCIDAE

The skink family is one of the largest groups of lizards with about 600 species and is almost cosmopolitan in distribution. Some 51 species in eight genera occur in South Africa. Some species are very common and often seen around homesteads. The Striped skink, *Trachylepis striata*, is one such species that has benefited from its association with humans. It has established itself widely and can commonly be seen on walls of houses and other structures throughout most of South Africa. Although not very colourful, the body is covered with smooth shiny scales, often brown with longitudinal stripes. This is especially so of the genus *Trachylepis*. The skinks live in a great range of habitats from mountain to desert, rocky outcrops, trees and many are terrestrial and fossorial. A burrowing mode of life leads to a great range of adaptations, mostly aimed at streamlining the lizard and reducing hindrances such as limbs and feet.

It is therefore in this family that the greatest range in limb reduction is found. The dwarf burrowing skinks of the genus *Scelotes* exhibit a remarkable range of limb degeneration. Some species such as the Montane burrowing skink, *Scelotes mirus*, have a full complement of limbs and digits although reduced in size; the Limpopo burrowing skink, *Scelotes limpopoensis* has four limbs, but digits on the forefeet are reduced to two or three. Other species only retain two hind limbs with varying degrees of digit reduction. Seventeen species have been described from South Africa of which 13 are endemic, many having very restricted distributions. The genus tends to occur mostly below the margins of the

The Limpopo dwarf burrowing skink showing limb reductions characteristic of this group of lizards.

The Giant legless skink is mostly found in forested areas and coastal bush.

TOP: *The head of the Giant legless skink.*
BOTTOM: *The Woodbush legless skink,* Acontophiops lineatus, *is an endemic of the forest margins in Limpopo Province.*

some live among the roots of grasses, and occur in a wide range of habitats from afromontane and temperate forest, savanna and grassland to arid sandy areas of Namaqualand, Limpopo Province and even mountainous areas fringing the Karoo.

The largest species is the Giant legless skink, *Acontias plumbeus*, which can grow to the exceptional length of 30 cm and can be 2 cm or more in diameter. It is a beautiful shiny purple black colour with smooth overlapping scales, and a large rounded scale on the snout, which covers half of the head. This assists the animal to break through the soil when burrowing. The tail is relatively short and stumpy so that reversal in the burrows is possible.

The genus *Acontophiops* is monotypic, that is it is only represented by a single endemic species, the Woodbush legless skink, *A. lineatus*, a species with a very restricted range in South Africa limited to the rocky grasslands fringing the afromontane forests of the Woodbush, near Haenertsburg, and in the Wolkberg. Although similar in appearance to lizards of the genus *Acontias* it differs in the number of bones forming the skull, and lies between these and the blind legless skinks of the genus *Typhlosaurus*, but is more closely related to the latter. This skink is also found under rocks and often in burrows, feeding on invertebrates such as termites. Like the other legless skinks, the species is live bearing, producing up to two young during midsummer.

The blind legless skinks of the genus *Typhlosaurus* appear to be the most advanced of the legless skinks. The eyes are vestigial, there are no external ears and the nostrils can be closed off by a plug, which is attached to the rear wall, all adaptations to a burrowing way of life. There are eight species and two subspecies occurring in South Africa. Like that of the dwarf burrowing skinks their distributions with three exceptions are coastal. All are psammophilous species mostly restricted to sandy soils through which they are able to burrow with ease. In fact, the distribution of sand is essential to these lizards. One species, the Striped blind legless skink, *T. lineatus*, is widely distributed in the Kalahari, often together with another

great escarpment from Namaqualand, along the coast to Zululand and Maputaland, extending inland into Mpumalanga and Limpopo provinces. They live in sandy or loose soil and lie up under rotting logs, stones and leaf litter usually buried in the soil. Many move by swimming through the sand with undulations of the body and tail. All species are live bearing, producing from one to four young.

South African lizard genera that have lost all signs of external limbs are *Acontias*, *Typhlosaurus* and *Acontophiops*. Most of the species in these genera are endemic to the country. The legless skinks of the genus *Acontias* are represented by eight species and five subspecies in South Africa. They are characterised by the large blunt nasal shield at the tip of the snout, cylindrical body and blunt-ended tails. These burrowing skinks are usually found in the soil under rocks and logs or in decaying plant material, although species, the Gariep blind legless skink, *T. gariepensis*. The sympatric occurrence of two species usually leads to competition for the same resources, namely food and shelter, which therefore leads to the exclusion of the one or the other. In this instance the former mostly inhabits the dune streets or valleys while the latter species occurs along the dunes. They therefore avoid direct competition for limited resources. The typical form of the Striped blind legless skink mostly occurs in areas of Kalahari sand of the Northern Cape Province, but there are two subspecies which are found widely separated from this in the Limpopo Province. These subspecies, *T. l. subtaeniatus* and *T. l. richardi*, occur in sandy terrain, the former in the vicinity of Vivo and the latter in Venda, east of Tshipise. During the last million years the climate of South Africa vacillated constantly between periods of high rainfall and very arid conditions. During the latter the sands of the Kalahari were dispersed by wind over much of the interior of South Africa, extending in the Limpopo Province probably as far as the north-eastern Kruger National Park and Sekhukuniland in the east. During this time species such as the Striped blind legless skink, Common barking gecko and the Serrated tortoise were able to extend their ranges eastwards. During ensuing wet cycles much of the sand was washed away leaving islands of sand where such arid adapted species continued to exist to the present time. However, through isolation these remnant populations developed characteristics of their own, but still resembling the original species to a large extent.

In a similar context, and possibly during the same climatic perturbations, the Golden blind legless skink, *T. aurantiacus*, and the Black-lined skink, *Trachylepis depressa*, were able to extend their ranges westwards from the coast of Mozambique to the north-eastern Kruger National Park in the case of the former and to the Nzhelele River in the case of the latter, when alluvial sands of the Mozambique plain extended well into the north-eastern parts of Limpopo Province. In the case of the former this also led to isolation and the development of specific characteristics.

Among the many peculiarities of this family is Sundevall's writhing skink, *Lygosoma sundevalli*, which is of considerable interest as it is a nocturnal species. It is also a burrowing lizard with reduced limbs and swims through the upper sandy soil layers by vigorous writhing actions. During the day it is chiefly found aestivating under stones or rotting logs. At night it becomes active and may move about above ground on its short legs with the elongate body undulating rapidly from side to side. It feeds mainly on termites and other small insects. In some parts of South Africa this unfortunate animal is called a 'springslang' because if molested it may shoot forward suddenly and is believed to be very poisonous. This is however not true and it is perfectly harmless and can scarcely bite. Its body is strongly muscled and covered by smooth scales so that it is held with difficulty. Like most

The Striped blind legless skink is generally associated with areas of Kalahari sand.

The so-called 'springslang', Sundevall's writhing skink, is a nocturnal lizard showing limb reduction, an adaptation to a burrowing existence.

lizards it loses its tail readily and 50 out 75 lizards examined showed evidence of caudal regeneration, indicating a substantial level of predation.

Two species of small skinks of the genus *Panaspis*, Wahlberg's snake-eyed skink, *P. wahlbergi*, and the Spotted-neck snake-eyed skink, *P. maculicollis*, occur in South Africa. Like geckos these lizards do not have moveable eyelids, instead the eyes are covered by a transparent brille permitting the skink to see at all times. Sexual dimorphism is present in these lizards with males developing a pinkish orange under the chin, throat and abdomen in the case of the former while males of the latter only have a pinkish colouration under the throat. Being small species with short legs, they forage for invertebrates especially termites in leaf litter at the base of shrubs and trees.

Most people are familiar with lizards of the genus *Trachylepis* as some species have adapted to living in suburban environments, especially the Striped and Speckled skinks, *T. striata* and *T. punctatissima* respectively. They are among a select few lizard species that have been able to withstand habitat destruction and alteration as a result of human activities and the impact of domestic animals, and have probably increased in numbers and in distribution. They are gregarious lizards often living in family groups and are commonly seen around houses, rockeries and on woodpiles. Live-bearing females produce from two to 10 young during midsummer with a second brood in late summer. Another interesting species is the Black-lined or Eastern coastal skink, *Trachylepis depressa*, which inhabits areas of deep sand in the north-eastern Limpopo Province, Mozambique and northern KwaZulu-Natal. It forages in open spaces between shrubs and grass tussocks but if pursued will suddenly dive and with swimming movements disappear into the sand to lie several centimetres below the surface. It forages mostly amongst leaf litter and between grass tussocks but at times also along the littoral zone at the edge of the sea.

While most species do not show a chromatic sexual dimorphism, one species, the Blue-tailed koppie skink, *Trachylepis margaritifer*, is characterised by differences in colour between mature males and females and juveniles. The former have orange-brown tails while the latter have blue ones. Juveniles in particular have brilliant blue tails but this tends to fade the older the animal becomes. Apart from tail colour, fully mature males are brownish black above and speckled or spotted with white, in contrast to the five longitudinal white stripes which extend from the head to the base of the tail in females and immature males. The latter may become as large as mature males but still retain a blue tail. Adult colouration probably only appears after two years of age, but it may also be linked to dominance.

These lizards are restricted to rocky outcrops and are particularly common on the large granite hills of the Limpopo and Mpumalanga provinces where they may be seen foraging and basking on boulders. Beyond our limits they occur in Zimbabwe and Mozambique. They are gregarious, living in family groups usually with a dominant male and often together with other lizard species in rock crevices. These skinks are oviparous, laying six to 10 eggs at a time during summer, usually under a rock on soil.

The Blue-tailed koppie skink male (top) and female (bottom).

Bouton's snake-eyed skink, a seafaring lizard.
Photo W.D. Haacke

Communal nesting takes place with up to 70 eggs having been recorded under a particularly favourable stone.

Skinks feed mostly on invertebrates but some species also include smaller lizards in their diet. Terrestrial and rupicolous forms seem to be fairly catholic in their choice of food. The burrowing forms are more specialised, feeding largely on termites but other invertebrates are also eaten. No South African species appear to feed on vegetable matter but several Australian skinks certainly are partly vegetarian.

Reproductively the skinks are a unique family with some species being egg-laying or oviparous, others giving live birth while at least two species may do both. The latter include the Cape skink, *Trachylepis capensis*, and the Variable skink, *T. varia*. The Variable skink, for instance, is mainly egg-laying in the northern provinces but is entirely live bearing in the Free State. Individuals from the Thabazimbi area and Mpumalanga Drakensberg escarpment are live bearing, which in the case of the former distribution gives rise to questions. It is normally assumed that live bearing is a survival strategy adopted under cold conditions while egg-laying appears to mostly present under warm conditions. In this respect the Cape skink appears to be egg-laying in the bushveld but live-bearing on the highveld, but individuals from the Alldays area are live bearing, confusing the issue! It is, of course, possible that such strategies may be a relict from the distant past, which has been retained into the present, as it did not negatively affect the survival of the species. There still remains considerable controversy as to which is more beneficial, live bearing or egg-laying under specific environmental conditions.

An American skink of the genus *Eumeces* exhibits some degree of parental care. In this case the female remains close to the eggs and throughout the incubation period turns them at regular intervals with her snout. No direct incubation takes place, as the animal cannot raise her body temperature.

One of the most interesting lizards of this family is Bouton's snake-eyed skink, *Cryptoblepharus boutoni*, so called because of the lack of eyelids, the eyes being covered by clear scales or spectacles. It is a small lizard, 10 to 12 cm long, with reduced limbs. It lives on the seashore on fossilised dunes and boulders, which may be partly submerged during high tide. It forages along the intertidal zone, running down the side of the boulders as the waters recede to catch tiny crustaceans that live in the tidal pools on the underside of the boulders which are exposed at low tide, as well as any that may have been washed up. A small population of this lizard occurs at Black Rock on the KwaZulu-Natal coast with the nearest other population near Inhambane in Mozambique. It is presumed to have rafted here on logs or other debris and established a colony. The population is small and relatively constant, numbering about 100 individuals, this probably being the number that the area can support. It was originally thought that all populations belonged to a single species with a wide distribution in the tropics, including Australia, and that the South African population originated in East Africa or Madagascar, but studies have shown that they belong to a group of species. Their final taxonomical status has as yet not been established.

The distribution of the family over the globe is very interesting. For instance North Africa has skinks that show affiliation to those of North America, while South America has members of the skink genus *Trachylepis*, which is common to Africa south of the Sahara. Similarly, ties with Australia are found in the genus *Lygosoma*, while links with Madagascar are maintained through the genera *Trachylepis*, *Acontias* and *Scelotes*.

THE LACERTID FAMILY
LACERTIDAE

The Lacertidae or lacertids are frequently referred to as typical lizards. They have scaly bodies, conical heads and most have moveable eyelids and normal, well-developed limbs, feet and tails. There are mostly no specialised body forms such as that of the skinks or other families discussed previously. They are fast running, can climb well, dig and even swim a little. There are about 250 species worldwide of which 27 occur in South Africa. They scarcely vary in body form, although some, such as the sandveld lizards, have very long tails that are as much as 1,5 times the length of the head and body. Almost all species in South Africa are terrestrial with a few that are rock living. These lizards occupy a wide range of habitats, from sandy desert to mountain tops, but avoid forested areas. Some species in Central Africa do inhabit forests, such as the Fringe-tailed lizard, *Holapsis*

TOP: *The Spotted sandveld lizard.*
BOTTOM: *The threat posture of Delalande's sandveld lizard, an uncommon montane and temperate grassland species. Photo Richard Boycott.*

guentheri, which has a row of projecting scales on each side of the tail that assist the lizard to climb up the boles of trees and act as a brace when it is on the trunk.

The terrestrial species have various adaptations for life on the ground, the most pronounced being those of desert-adapted species, which have elongated scales under the feet and toes to assist with traction on unstable surfaces, while in a number of species the lower eyelid has a large transparent disc in the centre through which the lizard can see even if the eyelid is closed. This brille is an adaptation to protect the eye from wind-blown sand and yet allows the lizard to see at the same time. These lizards are mostly small to medium sized, the sandveld lizards reaching a length of up to 30 cm. One species from south-western Europe and north-western Africa, the Jewelled lizard, *Lacerta lepida*, is the giant of the family, apparently reaching 75 cm in length.

Among some of the most beautiful of the lacertids are the sandveld lizards of the genus *Nucras*, a fact recognised by the ancient Egyptians who carried preserved mummified specimens as charms in small boxes.

Most species have a long pink or red tail and the body is spotted or striped particularly when young. One species has a blue tail and is extremely localised in South Africa only occurring in the Nwambiya sandveld of the north-eastern Kruger National Park. Most species live in open woodland and scrub, of which the Spotted and Holub's sandveld lizards, *Nucras intertexta* and *N. holubi* respectively, are amongst the most widespread, occurring over much of the savanna regions of South Africa. One species, Delalande's sandveld lizard, *Nucras lalandei*, is found in montane grassland from the Soutpansberg in the north to the Cape Province in the south and east, occurring down to sea level. It is probably the largest of the family Lacertidae in South Africa reaching a total length of 25 to 30 cm of which two-thirds comprises the tail. This attractive species has an olive green to brown body spotted with dark brown to black. The body is covered in very small scales.

Seldom is there such a difference in colour between juvenile (left) and adult (right) lizards as is found in the Bushveld lizard.

All species of sandveld lizards are diurnal and terrestrial, foraging for food among grass tussocks, slowly coursing the ground for prey but if disturbed rapidly taking shelter in the nearest bush. In very hot areas they may be seen during the hottest part of the day in the shade of shrubs, sometimes even climbing up into these bushes to a height of 30 to 60 cm above the ground. Here they may hang by their hind feet exposing as much of the body as possible in an effort to cool down as the ambient temperature approaches the animals' maximum tolerance level. They feed mainly on small beetles, grasshoppers, spiders and occasionally small scorpions.

Although the lacertids are unable to change colour there is change with age. This is especially well marked in the Bushveld lizard, *Heliobolus lugubris*, which in the juvenile state is black with yellow spots and stripes. As it grows older this contrast is gradually diminished and the adult lizard is brown above with paler longitudinal stripes and a white belly. It is interesting that this species occurs widespread in some of the hottest parts of South Africa such as the Kalahari and parts of Limpopo Province. With its black colour it would be expected to heat up considerably, but it forages in the open often almost until noon, before returning to the shade of shrubs and trees. It displays curious behaviour as it forages in the sun, running in short bursts of speed and then lying on its belly waving its forefeet about and lifting the hind feet off the ground. Similar behaviour recorded from other members of the family in the Namib Desert has proven to be linked to thermoregulation, the waving of the forefeet being an aid to heat dissipation.

The juvenile aposematic colouration in this lizard is also very interesting. As the animal moves along it periodically hunches its back but continues walking. Theory has it that this behaviour and colouration is mimicry of the Tiger beetle (Family Carabidae). This beetle is known to squirt formic acid from its abdomen if molested which can be painful and could cause temporary blindness should it enter the eyes. In mimicking this beetle it appears distasteful to potential predators, allowing more juveniles to mature.

Other species tend not to be as conspicuous and their young are mostly a nondescript brown, the colour of the soil. These lizards avoid the heat of the sun by moving in and out of the shade. Sexual dimorphism is present in some lacertids, the male being frequently larger, different in colouring or in the number of scales. Some species such as the Cape rough-scaled lizard, *Ichnotropis capensis*, exhibit such colour differences between male and female, the males tending to be more brightly coloured. This is usually associated with dominance, territoriality and reproduction. Usually the largest and brightest animals tend to be dominant. Lacertids usually forage by sight and are dependent thereon to recognise conspecifics and enemies.

Depending on where the colour is brightest the males threaten each other by a broad lateral display.

During September and October most lizards mate, like these Cape rough-scaled lizards, Ichnotropis capensis.

This is the case with the Cape rough-scaled lizard whose display consists of arching the back and flattening the body laterally, so as to present an imposing picture to his rival. The lizards circle each other in this manner until one of them takes the initiative and rushes in to bite his opponent and they wrestle about until the lesser of the two breaks away and runs off. However, actual fighting is reduced to a minimum, the imposing display and bright colours being sufficient deterrent.

Courtship is a simple affair among the lacertids, the male frequently approaching the female to touch her with his tongue, followed by jerky movements alongside her. If she does not dash off he may bite her in the neck after which he mounts her while she raises her tail. The male's position is to one side and one hind leg is placed across the female's back, while his tail slips under hers and he inserts a hemipenis into her cloaca. Both male and female may mate several times during the breeding season, which is mostly during summer, but not always with the same partner. There is no pair bond such as that in birds.

These lizards generally appear to be relatively short lived and some of them complete their life cycle within the space of one year. This is especially true of the Cape and Common rough-scaled lizards, *Ichnotropis squamulosa*, the former hatching from eggs between January and March and the latter during October and November, the eggs having overwintered after being laid in late summer. The hatchling Cape rough-scaled lizards develop gradually over winter until reaching sexual maturity during September and October and though they continue to grow, most adults die off during December with very few surviving to May. Six eggs are laid in a hole in the ground during November or December and incubate for eight to 12 weeks. In contrast the hatchling Common rough-scaled lizards only become mature later, at a time when most Cape rough-scaled lizards have already died. It has been suggested that this temporal separation is a result of competition for the same resources. By staggering the life cycles the two species may reduce competition for food.

These lizards feed primarily on termites, but spiders, grasshoppers and beetles are also eaten. One ambitious female Cape rough-scaled lizard was found dead with a large centipede stuck in her throat, and appears to have died from asphyxiation.

All lacertids lay eggs with the exception of a single species within this family. This is the Viviparous lizard, *Lacerta vivipara*, which is native to middle and northern Eurasia. Here it is an adaptation to climatic conditions as the species may be found within the Arctic Circle where temperatures even in summer are cool and the eggs would not incubate. The female can, however, maintain a higher body temperature by moving about in the sun. Where this species occurs in warmer climates it becomes egg-laying.

South Africa boasts some unique lacertids, including Smith's desert lizard, *Meroles ctenodactylus*, and the Wedge-snouted desert lizard, *M. cuneirostris*, both of which have wedge-shaped snouts adapted for diving into loose sand to escape danger and to retire at night. The former also has prominent serrated fringes on the toes assisting it to run on loose sand. Perhaps the most notable is the Shovel-snouted lizard, *Meroles anchietae*, from the Namib Desert. It also has the wedge-shaped snout enabling it to dive into loose sand and swim below the surface. However, this is not done solely to

TOP LEFT: An adult male Cape rough-scaled lizard, Ichnotropis capensis, *an 'annual' lizard.*

TOP RIGHT: The Shovel-snouted desert lizard – the raising of the hind limbs is a form of thermoregulation, heat being dissipated through the feet and the body is high off the ground. Photo: W.D. Haacke.

BOTTOM LEFT: The Shovel-snouted desert lizard, a sand-diving lizard of the Richtersveld and southern Namib.

escape predation but also enables the lizard to avoid excessively high temperatures by remaining buried under the sand during the hottest time of the day. The feet are also attenuated and fringed to enable it to run swiftly over an unstable substrate. Its thermoregulatory behaviour is very efficient, as it comes to the surface only when the temperature approaches 30 °C. It then basks pressing its body to the sand with all four legs held up in the air. As the air temperature rises to 40 °C it raises the body and straightens the legs. When walking it periodically raises the diagonally opposed limbs, using the tail as a support. The gait is stilt-like. This behaviour apparently assists in radiant and convective cooling. At surface temperatures above 40 °C it dives swiftly into the sand to cooler depths. A further interesting adaptation is the ability to absorb large quantities of water (up to 11 per cent of its own body weight). This is comparable with a 68 kg person drinking 6,75 litres of water. The water is stored in the colon of the hindgut. Lizards excrete literally solid uric acid and can therefore extract moisture for recycling. The Shovel-snouted lizard feeds on insects and plant seeds of which, when abundant, it will consume large quantities and then will not feed again for a prolonged period, remaining relatively dormant.

THE CORDYLID FAMILY
CORDYLIDAE

Throughout the eastern and southern parts of the savannas and steppes of Africa there occurs an interesting family of lizards – the Cordylidae, which includes the girdled lizards, flat lizards and grass lizards. There are two genera and 45 species, most of which are restricted to southern Africa and South Africa in particular. Probably the most familiar are the girdled lizards, of which the Sungazer or Giant girdled lizard, *Cordylus giganteus*, is the largest, reaching 30 cm in total length. Apart from the three grass lizards, it is the only truly terrestrial species, most others being rock living with a few partially arboreal. The Sungazer with its armour of heavy scales and spikes is also of special interest. It lives in small colonies, mostly inhabiting east-facing burrows up to 3 m in length and at its deepest about 45 cm below the surface. It digs its own burrows, which are characteristically oval with a ridge along the middle of the floor. Each burrow is usually occupied by a single animal, except when there are females with young. They are live bearing with one to two young but sometimes up to four usually born once every two years.

They are normally seen during the day sitting close to the entrance of their burrows, heads raised and bodies arched as they bask in the sun's rays. Here they sit and wait for passing prey such as beetles and grasshoppers; they usually do not forage far from their retreats. When threatened they retreat head first down the burrow and wedge themselves in at the deepest end by means of the large occipital spines while they thrash about with the very spiny tail, deterring any potential predator. They are preyed upon by snakes such as the Rinkhals, and birds such as Black-headed heron, the young being the most vulnerable.

Most of the girdled lizards – approximately 34 species and subspecies, of which 19 are endemic to South Africa – are rock living, occupying crevices between and under boulders. Both vertical and horizontal crevices are occupied, some species preferring the former and others the latter. All exhibit the tendency to wedge themselves in these crevices by arching the head and anchoring the occipital spines in hollows in the rock, while the spiny tail protects the more vulnerable sides and belly. One species, the Armadillo girdled lizard, *Cordylus cataphractus*, is exceptionally well armoured by spiny scales. It occurs throughout Namaqualand, living in small family groups inhabiting the same crevices in rocky outcrops. Although usually difficult to extract from these crevices, if it is dislodged it may bite its tail, forming a ball or hoop, presenting the spiny scales as a deterrent to a would-be predator.

The Giant girdled lizard or Sungazer, an endemic terrestrial lizard of the grasslands of south-eastern Mpumalanga and north-eastern Free State.

The Armadillo girdled lizard, a unique species from the rocky hills of Namaqualand and the western Karoo.

Most girdled lizards live singly but pairs occur, while they may form a colony in an isolated jumble of rocks. The Transvaal girdled lizard, *Cordylus vittifer*, usually inhabits small low-lying rocky outcrops, often at ground level, while other species such as the Dark and Warren's girdled lizards, *Cordylus warreni depressus* and *C. w. warreni* respectively, mostly require larger outcrops with boulders, often with sheltering vegetation scattered about, to provide shade and to camouflage these lizards when moving about. Here they may be seen basking close to the crevices or on top of boulders. They forage away from their crevices, dashing down to snap up passing prey and return as quickly to their retreat. The Dark girdled lizards occur along the Soutpansberg from east to west. Along their distribution range they change in colour from a relatively drab blackish colour in the east to a brightly marked black and yellow form in the west. This coincides with a similar change in rainfall, being very wet in the east at Entabeni and Thohoyandou and very dry along the Waterpoort and further west.

Three arboreal or partly arboreal girdled lizards are Jones' girdled lizard, *C. jonesi*, the Large-scaled girdled lizard, *C. macropholis*, and Tasman's girdled lizard, *C. tasmani*. The first species occupies holes in trees and dead wood, while the second occurs in dead wood and debris at the tide line and also retreat into *Euphorbia* plants. Tasman's girdled lizard occurs mainly under dead aloe leaves, those on the stem as well as leaves scattered on the ground.

The girdled lizards feed chiefly on beetles and other slow-moving prey which they grab in their jaws and chew with short hard bites of the lower jaw which is armed with conical teeth, the prey being swallowed almost whole. Only chitinous remnants are voided in the faeces, as everything else is digested.

All species of girdled lizards are viviparous giving live birth, producing from one to six young at a time. Births may be staggered with an interlude of up to seven days occurring between the first and second births. The young are fully independent following birth.

Very similar to the girdled lizards and formerly belonging to a separate genus *Pseudocordylus*, are the crag lizards. The latest revision indicates that they belong to the same genus as the girdled lizards. In general the crag lizards closely resemble some of the girdled lizards, with small scales on the back and spines confined to the tail only. These lizards occur mostly along mountain ranges and rocky hills, occupying crevices in rocky outcrops and along kranzes at higher altitudes than most girdled lizards, and rarely sympatric with the latter, which may be found on the same slopes but lower down. They are never found together on the same outcrop of rocks. There are five species and subspecies of these lizards, which are endemic to South Africa, Lesotho and Swaziland, some with wide distributions in South Africa.

Some species such as the Drakensberg crag lizard, *Cordylus melanotus*, display sexual dimorphism in colour and size with males being larger than females and having large blackish, triangular heads while the sides can be bright orange.

The males are usually seen basking on top of boulders in the morning sun, adopting an erect posture. They tend to occur in colonies usually with many females and juveniles. In behaviour and reproduction they are very similar to the girdled lizards with up to seven young being born.

The Cordylidae harbour some very interesting lizards which are found in the grasslands and savanna areas of the country. The grass lizards, as they are known, are slender lizards with a long whip-like tail. Three species occur in South Africa of which the Transvaal grass lizard, *Cordylus aenea*, is the only one with five digits on all four limbs. A second species has two toes per leg, while the third species, the Large-scaled grass lizard, *C. macrolepis*, has no forelimbs at all. Like that of other members of the genus the grass lizards are also live bearing.

Mention should be made of those jewels of the rocks, the flat lizards, genus *Platysaurus*, which exhibit a range of colours rivalled perhaps only by the dwarf chameleons. These lizards are exclusively rock living, inhabiting rocky outcrops, ridges and mountains. Here they live in narrow crevices emerging by day to bask,

TOP: *The Large-scaled grass lizard,* Cordylus macrolepis, *a unique adaptation for life in thick grass.*

CENTRE: *The Orange-throated flat lizard, an endemic rock-living lizard of Limpopo Province, is one of the many jewels among the reptile fauna of South Africa.*

BOTTOM: *A male Lebombo flat lizard, endemic to KwaZulu-Natal, Swaziland and adjacent Mozambique.*

forage or mate. They are specially adapted to living in these narrow confines as the head and body is flattened to enable the lizards to shelter in crevices no more than 5 mm wide. There are about 20 species and subspecies distributed from Mozambique and Malawi southwards to the Northern Cape Province, with the majority of species in Limpopo and Mpumalanga provinces and northwards. In fact, most species appear to occur in Limpopo where eight species and two subspecies are known. Three species and one subspecies are known from Mpumalanga and two each from KwaZulu-Natal and the Northern Cape respectively. Most of the flat lizards occurring in South Africa are endemic.

They forage and bask on rocks, usually only emerging when the rocks begin to heat up. They are often very active, leaping from rock to rock or running in short swift dashes as they forage over the rocks, snapping up their prey. Even when the rocks are so hot they almost burn the skin of one's hand these lizards may still be seen flitting about. Their food consists mostly of invertebrates such as beetles, ants and insect larvae, but two species feed largely on flower petals, young leaves and seeds.

The males are very brightly coloured, mostly in varying shades of green above with an orange-brown to pinkish-red tail. The undersides can vary from blue-green to azure to dark blue, while in some species it may even be black. Some Zimbabwean species have a purple belly while others have yellow spots. The females and juveniles are mostly much less colourful, being a dull grey black to brownish black above, usually with three white to off-white longitudinal stripes from head to base of tail. Many species also have white spots in the intervening spaces between the stripes on the back. Ventrally the females tend to be pinkish brown with dark spots in the corners of the scales.

Males appear to be territorial and have a number of females, juveniles and immature males living in the same area. These dominant males are exceptionally brightly coloured and pose on the rocks with head and body raised so as to show off the brightly coloured underside. They adopt this pose to attract females and to deter any other male of similar stature from entering their domain. Should this happen and another male does approach then the resident male will turn broadside and twist his body so that the ventral colours are exposed to the stranger. If this is still an insufficient deterrent then a fight ensues which usually terminates with the loser fleeing. Such fights appear to be quite frequent during the mating season, which is usually in early summer, the lizard torsos showing numerous bite marks at this time.

These lizards are oviparous or egg-laying, usually laying two eggs at a time under rocks. Several clutches may be laid in a season and some species may be communal nesters. The juveniles hatch in midsummer and reach sexual maturity in about their second or third year of age.

THE PLATED LIZARD FAMILY
GERRHOSAURIDAE

The plated lizards have on and off been incorporated in the Cordylidae but are currently considered to belong to a family of their own. The family is restricted to Africa and Madagascar but only three genera and 11 species occur in southern Africa. They range in size from the large Giant plated lizard, *Gerrhosaurus validus*, with a total length of 90 cm to the small Eastwood's plated lizard, *Tetradactylus eastwoodae*, which reaches 20 cm. The name 'plated lizard' arises from a skeleton of contiguous bony plates which lie under the epidermis encasing the body in a bony armour.

All species in this family are terrestrial although the Giant plated lizard lives in rocky outcrops. They inhabit grasslands, fynbos and savanna and may be locally common. The Five-toed plated lizard, *Tetradactylus africanus*, is quite common in the coastal grasslands of KwaZulu-Natal and the Eastern Cape provinces. Females of this species select the nests of an ant species *Anochetus faurei* in which to lay their eggs and several females may lay their eggs in a single nest. This appears to offer some protection from predation of the eggs. The ants also clean the eggs, which may prevent microbial

The five-toed plated lizard, a largely coastal lizard specially adapted to life in thick vegetation, with reduced limbs and a long snake-like tail. Photo: W.D. Haacke

attack, perhaps enhancing hatching success. It seems that the eggshell and hatchling cuticle contains similar hydrocarbons as those of the ants, which as a result do not attack the eggs or the young.

One of our rarest lizards belongs to this family. Eastwood's plated lizard, *Tetradactylus eastwoodae*, was originally described by the renowned herpetologist Dr V.F.M. FitzSimons from a specimen found near Haenertsburg in Limpopo Province in 1911. At that time the area comprised montane grassland, a typical habitat of these plated lizards. Subsequently the species was re-collected by Vincent Wager, just after the Second World War in 1949. Since that time the type locality has mostly been planted up with eucalyptus and pine forests and recent searches have failed to establish its continued existence. In fact, it has been listed as Extinct in the *Red Data Book on Reptiles and Amphibians*. This appears to be the first reptile in the history of South Africa to have gone extinct, a dubious distinction. Let us hope that it still exists in some remaining pockets of grassland. The practice of annually burning all remaining grassy areas for firebreaks may have resulted in the disappearance of this lizard. This practice continues to be to the detriment of the herpetofauna; most snakes found in such areas have healed burn marks down the back, having taken shelter under stones during winter, which is when most firebreaks are burnt, and having been partially scorched as a result.

Like most lizards, the plated lizards feed mostly on invertebrates. They vary in foraging strategies in that some species sit and wait for prey to pass while others tend to hunt actively, and are swift runners feeding on grasshoppers, beetles and anything else they can catch. One species, the Rough-scaled plated lizard, *Gerrhosaurus major*, is to a certain extent vegetarian and has been seen to feed on *Grewia* berries and some wildflowers. They live in burrows either dug by themselves or those dug by other animals. They are solitary except during the mating season, which is in summer. These lizards are egg-laying.

The Yellow-throated plated lizard, *Gerrhosaurus flavigularis*, shows a chromatic sexual dimorphism during summer with adult males having a yellow to orange throat. During the remainder of the year this colour may be absent. In certain areas such as the south-western North West Province and in the north-eastern Limpopo Province there are males with greyish blue throats. The possibility exists that these represent different species, although in lepidosis they appear similar. Like the other plated lizards these lizards are oviparous, laying nine or 10 soft-shelled oval eggs in

The Yellow-throated plated lizard, with its torpedo-shaped body is well adapted to rushing through thick vegetation.

The Dwarf plated lizard is a very attractive rupicolous species occurring from southern Angola south to the Little Karoo.

the soil or under stones. Incubation takes between 70 and 90 days.

A particularly attractive species is the Dwarf plated lizard, *Cordylosaurus subtessellatus*, which occurs on rocky outcrops along the arid west from the Karoo to Namaqualand and north through western Namibia to southern Angola. Dorsally the head and body is dark brown, flanked on either side by a creamy white stripe with dark brown along the sides, and a bright blue tail. It forages actively among rocks and associated vegetation searching for its invertebrate prey, but is quick to take refuge in narrow crevices between rocks.

A unique adaptation within this group is exhibited by two species, widely separated but both inhabiting arid areas. *Tracheloptychus madagascariensis* is a small lizard living in the coastal dunes of south-western Madagascar, around the town of Morondava. This lizard has a sharp, depressed snout and if chased dives into the loose sand and wriggles under with powerful twisting movements, disappearing several centimetres deep into the sand. The other species, the Namib plated lizard, *Angolosaurus skoogi*, also belongs to this family but is a much larger species with similar adaptations to a life in loose sand. It is interesting that this adaptation has recurred in several other lizard families such as the skinks and lacertids.

THE MONITOR LIZARD FAMILY
VARANIDAE

The monitor lizards or Varanids are conspicuous animals and the largest of the lizards. They are especially diverse in Australia where the smallest species is only 20 cm long, while the largest is the Komodo dragon, *Varanus komodoensus*, which occurs on three small islands and part of the island of Flores in Indonesia, and reaches a length of 3 m. They are the proverbial 'dragons' with a long head and neck, heavy body, thickly muscled tail and stout, sturdy legs armed with strong claws. The tongue is forked and extrudes snake-like as the animal moves along foraging for prey.

In South Africa only two species occur out of approximately 23 species, which are spread across the tropics of the Old World and especially Australia. At one time they also inhabited North America, as testified by fossil remains from about 60 million years ago. The two species common to most of South Africa are the Water or Nile monitor, *Varanus niloticus*, and the Veld monitor, *V. albigularis*.

The former may reach 2 m or more in length, while the latter is somewhat smaller, up to 1,5 m, but may become fairly bulky. The two species occupy different

Perhaps not as charismatic as an elephant, the Veld monitor is nevertheless a distinctive animal.

habitats, the former usually associated with water, while the latter is terrestrial and rock living. Both species are adept at climbing trees and rocks, but the Water monitor is adapted to water, swimming with legs adpressed to the sides and propelled by the side-to-side undulations of the tail. They tend to lie up along the banks of rivers or other water bodies, often on branches of trees overhanging the water. When frightened they scuttle off very rapidly and dive into the water, or slip from the branch into the water. They sometimes belly-flop from considerable heights. If not very alarmed they gently enter the water and swim out into deeper water and submerge only to reappear some distance away. They can stay under water for considerable periods, even up to an hour, without coming up for air but mostly resurface much sooner and with greater frequency. The Veld monitor, in contrast, frequently resorts to climbing trees, particularly when chased. If pursued up the tree the animal will drop from the end of a branch to land flat on its belly only to leap up again and run for a more permanent shelter. It is a very hardy animal.

Monitors are predacious, feeding on almost anything they can overpower. The Water monitor consumes mainly frogs and crabs, but birds, eggs, lizards, small mammals and invertebrates are also important sources of food. It can be a menace to poultry, especially feeding on eggs and young chickens. At the same time it may also scavenge on carcasses killed by other animals. The Veld monitor may also make raids on chicken coops, but feeds largely on invertebrates, particularly slow-moving beetles such as Toktokkie or Darkling beetles and millipedes which are easy to catch. They are also opportunistic and eat birds, snakes, including poisonous species, and small mammals. The prey is usually swallowed whole and not chewed for any length of time.

Water monitors occur along most rivers and dams in South Africa except in the southern Cape.

A Veld monitor with inflated throat, body held high and arched tail, ready to defend itself. Photo: Lorna Stanton

The Water monitor is well known for its habit of locating crocodile nests, digging them up and eating the eggs and will even overpower and eat hatchling crocodiles.

It displays considerable cunning in attempting to approach a nest guarded by a female crocodile. A world-renowned herpetologist Captain C.R.S. Pitman recorded how a pair of Water monitors approached a female crocodile guarding her eggs. The one enticed the crocodile to chase it, whereupon the other dug up the eggs, started feeding and was soon joined by the other monitor, who had eluded his pursuer. They finished off a good portion of the eggs before the female crocodile returned to the nest. Another observer saw a Water monitor uncover a nest of crocodile eggs and carry them off one by one into the adjoining reeds, where it crushed the eggs and fed on the contents. The shells were discarded and the lizard then returned to the nest for more.

Monitor lizards appear to mate in early spring. Although little is known of its social life, in some species the males have been observed to fight in a strange and spectacular fashion, rearing up on their hind legs and grappling with each other with their powerful claws. All species lay eggs, the South African species during November and December. At this time a Veld monitor may dig holes up to 40 cm deep. The eggs are oval and blunt-ended, numbering 10 or more. The soil is raked back into the nest and eggs covered and tamped down with the snout. During a study of the Water monitor by Raymond Cowles in Natal, he found that the eggs were laid inside living termitaria. They took 10 months to incubate, the young digging out of the nest on hatching. Eggs incubated under semi-natural conditions hatched in five to six months. It is possible under natural conditions that the eggs of both species may take as much as 10 months to incubate as the eggs may overwinter only hatching in the following spring.

Both species are diurnal and may frequently be seen basking on rocks or river banks during the morning. When a pool dries up the Water monitor may move overland to another one, often travelling extensive distances across country. The Veld monitor also moves about extensively and many are killed while crossing roads. It is likely that they occupy discreet home ranges, although this has not been confirmed to date. Tagged lizards have been recorded from the same area over a period of two years.

On account of their large size they have few enemies, amongst the most important are the Martial eagle, *Polemaetus bellicosus*, Brown hyena, *Hyaena brunnea*, and of course humans. There are records of pythons feeding on monitors, while crocodiles will also devour them, especially females protecting their eggs. If a monitor lizard is cornered it fights back and distends its body amidst loud hissing and standing on straightened legs thereby appearing twice its normal size. With its mouth open and the throat puffed out, it lashes out violently with its hard leathery tail. It may also lunge forward to bite and the crushing power of the jaws can inflict a very painful bite. The teeth, however, are relatively small and round cusped, but the jaws clamp strongly on the aggressor. While biting it may also rake the aggressor

Being unobtrusive is the key to survival, a Veld monitor resting amongst the branches of a fallen dead tree.

with its strong curved claws. If this show of force is not successful, the animal, especially the Veld monitor, may play dead and even allow itself to savaged by dogs, but when the aggressors tire and move off, it will slowly roll over again and gradually move off, at first slowly, but with ever increasing speed until it feels safe.

Monitors are especially prone to tick infestations, which can be a problem. The males of the tick genus *Amblyomma* can be found crammed in the nostrils and even in the nasal passages. The females of this tick seem to prefer other parts of the body, such as the junctions of the limbs and body or around the cloaca, but are far fewer in number. The congestion of ticks in the nostrils has been known to cause death in animals with abnormally heavy infestations.

AMPHISBAENIANS
SUBORDER AMPHISBAENIA

THE WORM LIZARD FAMILY
AMPHISBAENIDAE

The amphisbaenians are a unique and curious group of reptiles, which are mostly adapted to a burrowing and subterranean existence. There are some 150 species widespread in Africa, Asia, Europe and the Americas. Originally they were classified under the lizards, but most recently have been placed in a suborder of their own, the Amphisbaenia. Almost all species are limbless with one exception from North America, which boasts two forelimbs. Because these animals are pink and cylindrical in shape, and have the body subdivided into circular rings, called annuli, they resemble earthworms, hence the name worm lizards.

It has been proven that they originated during the Mesozoic era, some 200 million years ago, at which time they became adapted to a burrowing mode of life. This is especially evident in the genus *Monopeltis* in which the scales of the head have been superseded by an enlargement of the rostral scale, which covers the top of the head. This fingernail-like shield is remarkably shaped to facilitate burrowing, enabling it to force its way through the soil. The eyes are vestigial, although still visible externally as a dark spot, and probably only able to discern between light and dark. Another notable difference from most lizards is that the tongue cannot be withdrawn into a sheath and the teeth are few but may be relatively large. Caudal autotomy is present in the worm lizards. In some species, such as the Msimbiti worm lizard, *Chirindia langi*, the tail exhibits a distinct line approximately one third of its length. This line demarcates the caudal fracture line, along which it breaks off when seized by a predator, but contrary to most lizards it does not grow again.

In contrast to most lizards and snakes, which either have two lungs or the left lung reduced, the amphisbaenians have only one lung, as the right lung is reduced.

The name Amphisbaenia originates from Greek and Roman mythology, where it referred to the fabled serpent with a head at each end, enabling it to move both ways. Many snakes are able to do this also, but usually in a sinuous manner and not solely using the body musculature. It is an adaptation to a burrowing existence and some legless lizards are also able to do this. Other related adaptations consist of the degree of ossification and overlap of the bones in the skull, which instead of abutting at their edges are more crowded.

Of the approximately 24 genera and 150 species found throughout the world, only four genera, nine species and three subspecies occur in South Africa,

TOP LEFT: The Dusky spade-snouted worm lizard, Monopeltis infuscatus, *is a fossorial burrowing reptile, mostly limited to a broad belt from the Kalahari to the lowveld of Limpopo Province.*

TOP RIGHT: The Msimbiti worm lizard, an endemic worm lizard from the north-eastern Limpopo Province, is usually found under stones and rotting logs.

BOTTOM LEFT: The Kalahari round-headed worm lizard tends to occur mostly under stones and rotting logs.

where some have wide distributions, while others are more restricted. It appears that the species are more or less evenly distributed across the country, and most species occur in savanna woodland or scrub. Although some correlation can be made with soil type and structure, it is difficult to generalise. On account of their mode of life they are rarely seen on the soil surface with some exceptions during the rainy season and when areas are cleared during road-building operations, especially when virgin soil is ploughed over.

A recent revision of the genus *Monopeltis* has subdivided the South African spade-snouted worm lizards into five species, the most widespread of which is the Dusky spade-snouted worm lizard, *M. infuscatus*. As mentioned before the very depressed, sloping spade-like snout is ideal for burrowing even in relatively hard sandy soil. Some species may have considerable tunnel systems while others make do with much less. They are rarely found under objects but are often exposed at road construction sites when the top 10 cm of soil is removed. In softer soils burrows may go as deep as 30 cm, as for instance at Ndumu in Maputaland.

Other species include the Blunt-tailed worm lizard, *Dalophia pistillum*, the Msimbiti worm lizard, *Chirindia langi*, and the round-headed worm lizards of the genus *Zygaspis*. The Blunt-tailed worm lizard prefers sandy soils and lies shallowly in the first 10 cm of soil, often under decaying vegetation. This very rare species has only been recorded from the Waterberg of Limpopo Province and parts of the Northern Cape. Two species of the round-headed worm lizards occur in South Africa, the most common being the Kalahari round-headed worm lizard, *Zygaspis quadrifrons*, a Kalahari species that is most often found singly under rocks, rotting logs and occasionally under leaf litter lying shallowly in the soil.

All amphisbaenians are pink, pinkish brown or brownish purple. The Msimbiti worm lizard is pink and translucent on the underside and is very small, not exceeding a length of 17 cm, whereas some of the larger spade-snouted species may reach 40 cm.

The Msimbiti worm lizard is a very interesting and localised species, getting its common name from the fact that it is mostly restricted to areas in the Soutpansberg where the Msimbiti or Lebombo ironwood, *Androstachys johnsonii*, is found in pure stands along the mountain crests which can be discerned from afar. However, this worm lizard also marginally occurs in open Mopane woodland. No other vegetation except some grasses grow here and the terrain is usually rocky. Its association with this vegetation peters out to the west of Wylies Poort and at Mara it may be found in open bushveld with baobabs, knobthorns and bushwillows. However, the worm lizard occurring here is a subspecies of the typical race, which may explain the more catholic choice of habitat.

These worm lizards seem mostly to occur under rocks in moist soil, but are also found under rotting logs, where they either lie on top of the soil or inside small burrows or tunnels, usually with the rock as the roof. They appears to be sensitive to sunlight, perhaps due to the lack of pigmentation and try to hide almost immediately if the rock is lifted. The soils here are mostly loam or sandy loam but occasionally these worm lizards may live in areas of clayey soil. They are usually solitary, but occasionally more than one individual and sometimes up to five may be found under a rotting log. They may be concentrated in favourable sites, on one occasion nine worm lizards were recorded from an area of roughly 20 m^2, some under adjacent rocks, others more dispersed. They feed largely on termites and have small conical teeth with which they chew their prey.

Little is known of the breeding behaviour of the amphisbaenians. Records indicate that they are oviparous or egg-laying, laying up to five eggs, but as the ecology of most species is unknown it is difficult to generalise. They are preyed upon by birds and mammals especially mongooses which with their keen sense of smell are able to locate them. They are also preyed upon by several species of burrowing snakes, particularly quill-snouted and purple-glossed snakes. It has been suggested that each species of the former feeds on a specific species of worm lizard, but this appears to be pure conjecture. While our knowledge of these animals remains poor, we can but hypothesise about the life and habits of these enigmatic reptiles.

Both aquatic and terrestrial, the Green water snake rarely climbs into shrubs or small trees. Photo: J. Marais

SNAKES

ORDER SQUAMATA

SUBORDER SERPENTES

Snakes in general are probably the most renowned of reptiles. While lizards may easily escape notice or are rarely given a second glance, the reaction to snakes is the dramatic opposite. It is little realised, let alone believed, that snakes are very timid animals which would rather be left alone to continue in pursuit of their own survival. It is largely agreed that snakes originated from some lizard-like ancestor possibly belonging to the Varanidae or the related Lanthonotidae. Although none of these lizards are limbless or show reduction in limbs or limb size, various primitive characters lend support to this idea. Many other lizards of different families have reduced limbs, digits and even the complete loss of limb. This development is mainly the result of parallel evolution or convergence, which was discussed earlier.

Fossil remnants of snakes are not common, probably due to the fact that their bones are relatively thin and easily decomposed. However, remains have been found which date back as far as the Cretaceous era, when snakes were rather short and thick bodied and possessing both lizard and snake characteristics.

ADAPTATION

Although snakes are mostly considered to represent the 'typical reptile', they are in fact highly specialised animals, which have achieved perfection in a limbless way of life. They have adapted to as wide a range of habitats as the lizards. They can be found in sandy deserts and in the oceans, while their ranges extend almost to the Arctic Circle and to altitudes of up to 6 000 m. They include specialised burrowers, climbers, swimmers and display an ease of movement unparalleled among vertebrates. They are extremely agile and eat an incredible range of food, which has contributed much to their success.

The versatility of snakes is well illustrated by the fact that they are often found in the vicinity of human environments, be it in barns in rural areas or in urban gardens. Such commensals include the Brown house snake, *Lamprophis fuliginosus*, Herald or Red-lipped snake, *Crotaphopeltis hotamboeia*, and Common egg-eater, *Dasypeltis scabra*. Their survival under these conditions is due to the establishment of refuges in rockeries, building rubble and walls, as well as their nocturnal habits. There is always a plentiful supply of food in the form of lizards, mice and toads, which are likewise attracted to human habitations on account of the food and shelter available.

Approximately 2 600 species in more than 420 genera occur worldwide with the exception of the Poles. Of these, 108 species in 41 genera occur in South Africa.

Similar to lizards, snakes are most plentiful in the tropics and the number of species decreases with distance from the equator. The northern provinces and KwaZulu-Natal have the greatest variety of species in South Africa and on a typical bushveld farm as many as 29 snake species may be recorded. It may be of interest to split them up according to the type of habitat that they occupy. For instance it may include six burrowing, 19 terrestrial and four arboreal species. Among the

One of the commonest snakes and amongst the most widespread, the Brown house snake can be considered a human commensal.

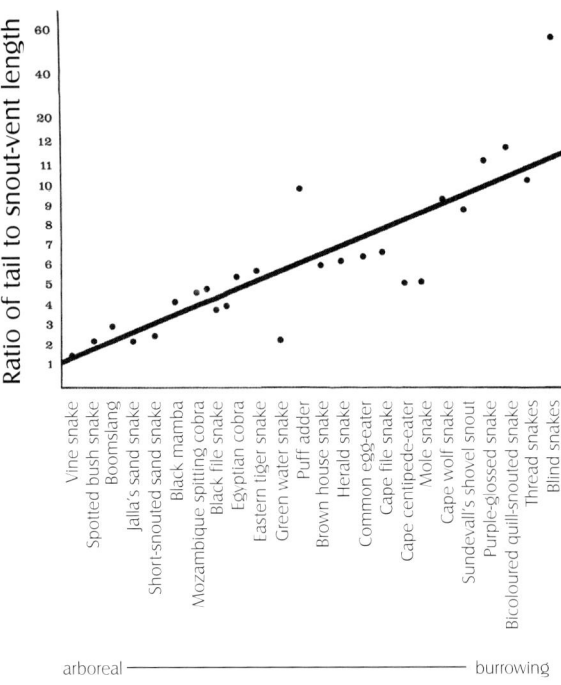

Ratio of tail length to head and body length in different South African snake species.

burrowers there are some that are exclusive burrowers as well as those that tend more towards a terrestrial existence. Similarly, among the terrestrial species there may be transitions to a burrowing lifestyle. On the other extreme some may be transitional to an arboreal existence and amongst arboreal species there may be a tendency towards terrestrial habits. These habitat choices are reflected in the ratio of tail length to snout-vent length. A large ratio indicates a burrowing mode of life while a small ratio is more characteristic of an arboreal existence. A long tail is a distinct disadvantage when trying to reverse in a burrow, while it is advantageous to a tree-living species, in which case the tail needs to be agile and capable of assisting the animal to cross between branches. Most sand snakes, which are terrestrial, also exhibit small ratios as they are racers and actively pursue prey. Their main prey is lizards and mice, which are able to move fast to avoid predation. The tail acts to propel and balance the snake as it chases its prey. The Puff adder, *Bitis arietans*, has a high ratio, being a slow-moving snake which catches its prey by lying in wait at a rodent pathway. It therefore does not rely on agility to catch its prey, instead waiting for it come close and catching it with a lightning quick strike. Each species has therefore become well adapted to make best use of its specific niche.

ANATOMY

The suborder Serpentes is characterised by the lack of limbs or girdles, except in a few forms where these are relict. The body is greatly elongated and the tail is of varying lengths, depending on mode of life and habitat. The vertebrae have additional facets to allow for greater articulation, which facilitates the serpentine type of movement. Furthermore there are no caudal fracture planes as in lizards but some snakes do twist their tails off, though no regeneration occurs.

The lower jaw consists of two bones joined by an elastic ligament. This is flexible to facilitate feeding and the tongue can be completely withdrawn into a sheath in the floor of the mouth. The palatal bones are moveably attached to the skull and the total number of bones forming the skull has been reduced. The parietals are always fused and together with the frontals have descending processes, which completely enclose the brain case. The teeth are recurved, very sharp and brittle and are continuously replaced. There are usually four rows of teeth in the upper jaw situated on the maxilla, palatine, pterygoid and dentary bones. The teeth are of the acrodont type.

The body is covered with overlapping, non-granular scales and the eyelids have fused to form a transparent spectacle over the eye. There is no nictitating membrane, as occurs in most lizards and there is no external ear opening, tympanic cavity or Eustachian tube.

Most snakes only have one lung, normally the right one, while the left side is almost vestigial. However, some variations do occur. The probable reason for the single lung, which can extend to two thirds of the body, can be attributed to the streamlined body, caused by the degeneration of the limbs, so that there is only enough space for one lung to function properly. Other paired organs such as the kidneys and testes are staggered at different levels in the body.

Thermoregulation

Snakes are also 'cold-blooded' animals relying on the sun to create conditions suitable for movement, prey capture and reproduction. In South Africa there is a pronounced seasonality to snake activity, which follows that of the lizards and amphibians. During the period mid-May to mid-September snake activity is virtually brought to a standstill. This may relate to their inability to capture and digest food at this time. It may also be due to an inability to digest food at lower temperatures.

Most snakes tend to be relatively inactive at this time but some may move about, emerging from their refuges to bask in the winter sun, but rarely feeding. Most snakes retire to holes in the ground or in trees, logs, under stones, in termitaria and in clumps of grass or other vegetation. Even piles of building rubble provide adequate refuge. In one such heap for instance 21 Striped skaapstekers and one Common night adder were found, each occupying a separate cavity, and not in contact with one another. Sustained temperatures of 4 °C for a few hours could kill a snake, so that it generally hibernates where ambient temperatures remain above this level. Soil temperatures become relatively stable at a vertical depth of 15 cm below the surface, which makes burrows in the soil important refuges for reptiles. Snakes caught outside during a cold night become very sluggish and lethargic and may appear to be dead but when warmed up gradually can be revived. From mid-September onwards snake activity increases with an increase in the minimum temperature above 10 °C, most species mating during October although some species such as the Puff adder may start earlier. Warm weather is essential for this. While snakes are active during the day, they retire to shade at midday if it becomes too hot, practising similar shuttle behaviour to that of lizards.

Many snakes are, however, nocturnal and rarely see the sun. However, the ambient temperature is high enough to enable such species, which may have a lower thermal lability, to be active and in a condition to digest their food. Most nocturnal species tend to be slow moving and night temperatures are sufficient for foraging and feeding.

Food

Food and feeding are as varied and diverse as the snakes themselves. Although the majority feed on lizards, frogs, toads, birds and mice, there are others which take birds' eggs, lizard and snake eggs, bats, snakes, centipedes, termites and snails. They capture their prey in various ways, most hunting by sight, reacting to movement by prey animals with a swift dash and bite. Some species may follow the scent trails of prey. Some will bite and throw its coils about the prey, constricting it so that the animal dies from asphyxiation and cardiac arrest. Only some snakes are poisonous, most injecting their poison by means of specialised elongate teeth or fangs, which are either located at the front of the mouth (the so-called front-fanged snakes) or just below or behind the eye (back-fanged snakes). The poison is injected either by a direct bite as applied by the front-fanged species or else the prey is manipulated towards the rear of the mouth and the poison chewed in, as is the case in back-fanged species. The latter species feed on prey that is unlikely to injure the snake during the process of moving the prey towards the poison fangs. When the poison has taken effect, the prey relaxes and is swallowed whole. Prey that may harm the snake, such as mice, are usually bitten and released. After a while the snake follows the trail of the prey by which time the poison has taken effect and the prey is no longer in a position to defend itself and is eaten without harm to the snake.

The poison is manufactured by the salivary glands, which are contracted by muscles that force the poison via ducts to canals inside or along grooves on the front of the fangs, where it is injected. It appears that some non-fanged snakes also have poisonous saliva, which they chew in when the snake bites its prey. Contrary to popular belief, the tongue, even in poisonous snakes, is quite harmless and is used as a sense organ, in conjunction with the organ of Jacobson, to follow prey or to investigate it before swallowing. The tongue is also used in mate recognition, as well as to familiarise the individual with its surroundings. Once the prey is immobilised the snake releases its hold and proceeds to nose it, flicking its tongue along the body until it

A Spotted house snake in the process of constricting a lizard.

orientates itself to the position of the head and body. Small prey can be swallowed from any angle, but larger prey is always taken head first, so that the limbs lie adpressed to the body to facilitate swallowing. The latter is assisted by the fact that the jaws of snakes are moveable, being loosely attached to the rest of the skull by elastic ligaments. Most of the bones of the jaw and palate are armed with small, sharp recurved teeth, which assist in manipulating prey and drawing the food into the jaws. The halves of the lower jaw are united in front only by elastic tissue so that they can move separately in conjunction with the two sides of the upper jaw to draw prey into the mouth. The brain case, on the other hand, is strong and compact to protect it from possible injury. In addition, there is a copious flow of saliva, which facilitates swallowing. The skin of the neck is also distensible, while the windpipe or glottis can be protruded from the throat, so that breathing is not obstructed when large prey passes down the throat.

Once the prey has been ingested the snake retires to a secluded spot to digest the food. This usually takes four to five days, but is of course dependent on the size of the prey and the ambient temperature during this time. Feeding during summer months may be on a regular basis, but little or no food is eaten during winter. Snakes can easily survive without food for four months at this time of the year, due to low temperatures and resulting low metabolic rate. A Southern African python, *Python natalensis*, once refused food for a period of more than two years, though shorter periods of time are more common. During this time the animal lives off the large fat reserves situated on either side of the body along the intestines.

HEARING

When wandering about the veld in summer it is surprising how few snakes are actually seen. This could indicate a paucity of snakes but could also be due to their cryptic colouration, which makes them difficult to detect, or their flight when aware of someone approaching. Although these aspects all have their merits, it is probably the latter that contributes most to their invisibility. Snakes have no eardrums, as the stapes or stirrup bone is connected to the quadrate instead of the eardrum as in most other reptiles. It is therefore unlikely that snakes can hear airborne sounds. They do, however, appear to be sensitive to ground vibrations, which are transmitted through the bones of the jaw and skull to the inner ear. They can therefore perceive footsteps and other disturbances causing them to move away or into cover. The well-known snake charmer with his flute does not charm the snakes into moving with his music; instead the snakes adopt a defensive posture and follow his movements as he sways from side to side.

Behaviour

While lizards exhibit behaviour patterns which to some extent parallel that of mammals, snakes do not appear to have such stereotyped behaviour. There is still much to be learned about their social life and behaviour. Snakes in general appear to be solitary animals. Some species, such as rattlesnakes and garter snakes of North America, are more or less gregarious and may congregate in the spring mating season or before hibernation in autumn. This also appears to apply to local species such as the Vine snake, *Thelotornis capensis*, and Striped skaapsteker, *Psammophylax tritaeniatus*, as well as a few others. Some species hibernate communally but disperse once weather conditions become favourable – this applies to some rattlesnake species, and although reports of such clanning have been received with regard to the Cape cobra, *Naja nivea*, this has not yet been confirmed. Snakes rarely come into contact with others of the same species except during the mating season. A few snake species may form some type of 'pair bond' in that two and on rare occasions even three animals may be found inhabiting the same hole. The Black mamba, *Dendroaspis polylepis*, is one such species and in suitable habitats may live like this for years. Richard Isemonger, in his book *Snakes and snake-catching in Southern Africa*, records some observations in this regards. However, little social contact is made even under these conditions with the exception of a flick of the tongue in recognition or when basking together at a favourable site.

Reproduction

During the mating season, which for most snakes starts during the latter half of September, they behave quite differently, particularly when opposite sexes meet. The male tends to approach the female in a jerky manner, accompanied by a similar spasmodic side-to-side darting of the head. He noses the female along the length of her body, rubbing or touching her while his tongue flickers rapidly in and out. After some stimulation the female arches her tail and the male twists his tail under it so that their cloacas come in contact. Copulation has to take place as snakes fertilise internally. Similar to lizards, males have two structures called hemipenes which are normally kept inverted in the base of the tail. The sex of many snakes can be determined by the swelling at the base of the tail in males and the absence thereof in females. Also in most species males have longer tails than females. When copulating the male everts one of the hemipenes and inserts it into the cloaca of the female and sperm is transported down a groove along one side of the hemipenis. There is considerable variation in duration of mating but this usually lasts from a few hours to as long as a day or even longer. Throughout this time the animals lie entwined and due to the locking mechanism in the hemipenis cannot be separated without injury to the participants.

After fertilisation the eggs develop and are ready to be laid some six to eight weeks later. The snakes choose a variety of sites, usually in a position where the soil is moist but not wet, and where temperature extremes are minimised. Preferred sites are under rocks on soil, in

The Striped skaapsteker is one of few South African species which has been known to form aggregations.

holes and in rotting vegetation. The eggs are laid fairly rapidly, within a space of two hours, but can take up to a day or more in some species with large clutches. A python, for instance, may lay up to 50 eggs, taking one or more days to complete this. Once the eggs are laid they are with a few exceptions left to incubate on their own. In South Africa only the python is known to incubate her eggs. There is considerable mortality as dehydration, fungi and other animals take their toll. The eggs are subjected to periodic bouts of wet and dry conditions, which affect hatching success. The hatchlings break out of the egg, using an egg tooth on the tip of the snout. They do not emerge immediately but spend some time with only the tip of the snout appearing at the opening. Eventually they make their way out as miniature replicas of the adult. Residues from the egg may assist in moistening the surroundings enabling them to make their way to the surface. If unable to do so they will die.

Not all snakes are egg-laying. Many are live bearing or viviparous and some are ovoviviparous. The former give birth to live young while in the latter the eggs are retained in the oviducts of the female and are laid shortly before the young are due to emerge. This may still take from a few hours to several days, however the eggs are always laid when the young are partially to fully developed, but still enclosed by the shell.

There are distinct advantages to viviparity as opposed to egg-laying, as eggs are greatly influenced by the vagaries of the climate, depending on whether the rains may be early or late and whether thermal conditions are good or bad. The type of soil, where the eggs are laid, what depth and a host of other variables also play a role in successful incubation. There are many predators of eggs such as mongooses, pigs and other snakes, while ants also take their toll. Live-born young are at an advantage as they are almost self-sufficient immediately after birth, and the female can select a secluded site in which to give birth. Live birth is particularly advantageous in areas with a cool climate where ambient temperatures remain relatively low, so that incubation would be drawn-out perhaps taking twice as long or longer for the eggs to hatch. In comparison a gravid

The eggs of the Spotted skaapsteker with one of the females around several clutches.

female can move around following the sun and sustain a higher body temperature while basking, ensuring that the young develop under optimum conditions. Disadvantages associated with viviparity are that the female has to provide sufficient food to nourish the developing young for many weeks, while at the same time the female becomes less mobile and cumbersome and therefore less able to avoid predation. Ovoviviparous species are intermediate between these two, as the developing embryo is nourished by the egg yolk and at the same time it is still carried within the body of the female. All three methods of incubation can be found in South African snakes, and will be discussed under their respective families.

LOCOMOTION

The swiftness of snakes on the ground and in trees has been the subject of much speculation and story telling. In particular species such as the Black mamba have given rise to many hair-raising episodes. Snakes basically move in two ways. The first, which is normally associated with these animals, is a side-to-side undulating movement whereby the snake throws its body into a series of lateral waves that flow continuously from the head to the tail. With this movement the sides of the body are thrust against projections and irregularities on the ground, such as vegetation and stones and thus seem to propel it effortlessly along. However, it moves with great difficulty over relatively smooth surfaces such as tarmac or wide sandy roads. This can be seen in the wide tracks made by the snake on a sandy surface, which suggest that a much larger animal had crossed over. The short distances between the curves, however, point to a much smaller animal.

The second method of locomotion is called rectilinear locomotion, the name derived from the fact that the snake progresses in a relatively straight line using the broad ventral or belly scales which are independently activated, each one being successively raised and lowered by small muscles attached to the ribs. This caterpillar- or millipede-like locomotion is particular to large-bodied snakes such as pythons, thick-bodied vipers and adders. During locomotion the broad scales hook onto projections and prevent the snake from slipping back. None of the more slender species moves by this type of locomotion, although these scales are also important in other terrestrial and arboreal snakes. Some arboreal species such as the Spotted bush snake, *Philothamnus semivariegatus*, have keeled ventral scales which enable them to climb up the boles of trees or the unplastered walls of houses, provided there are enough projections.

The speed of snakes has been greatly overrated and few species can even achieve more than 20 km/h. Most species are much slower at a speed of 4 to 6 km/h. The speed of the fabled Black mamba lies in between these two and the animals cannot by any stretch of the imagination move as fast as a galloping horse.

Diseases and enemies

Snakes are prone to diseases and parasites. Ticks are commonly found attached to the skin between the scales where it is not protected. Even aquatic species are known to have ticks, the most common being the snake tick. Mites are also present and can cause considerable discomfort, debilitation and even death. Other insects known to feed on snakes include mosquitoes and two species of Tsetse flies.

Internal parasites are numerous and include various amoeba-like species, which can cause death, while nematodes and cestodes are also common, and so are large worm-like parasites called pentastomes. Trematodes are also abundant in certain areas occurring in the buccal cavity. Some mites are also found internally in the lungs. Blood parasites include trypanosomes, plasmodium and haemogregarines, which are protozoans living in the erythrocytes of the snakes.

Snakes also suffer considerably from trauma, both physical and emotional, particularly after capture, although this is also known from the wild state. Necrotic stomatitis (canker or mouth rot) and necrotic dermatitis (scale rot) are responsible for many deaths, even in the wild state, but especially in captivity.

Snakes have many enemies, foremost of which are humans, who destroy ruthlessly and recklessly due to fear, ignorance or religious beliefs, or for skins, food and medicine. The situation is exacerbated by habitat destruction through agriculture, urbanisation, mining, afforestation and road construction. One need only to travel country roads during summer to see the numbers of snakes killed on roads. Natural predators are many and varied and include many birds of prey, two of which are called snake eagles. The Secretary bird will eat anything in its path. Many other birds such as storks and herons, rollers, kingfishers and shrikes also feed on small snakes, including the young of larger species.

Many mammals feed on snakes including jackals, wild cats, genets, civets, mongooses and even leopards, which may kill a python. Mongooses are generally considered snake killers but these animals mostly eat invertebrates, but will take whatever they come across, including lizards and small snakes. Large snakes are rarely killed, although the Slender mongoose, *Galarella sanguinea*, may kill Vine snakes. However, large numbers of small snakes fall prey

Banded mongoose, hooligans of the veld and predators of smaller snakes.

to mongooses. Different mongoose species apply different tactics when killing medium-sized snakes. Slender mongooses, due to their solitary way of life, depend on their speed and agility as well as endurance when attacking a snake. They adopt a series of sham attacks by leaping forward and backward, eventually tiring the snake, and finally leap on it biting it behind the head and severing the vertebrae. The social Banded mongooses, *Mungos mungo*, attack the snake in a pack and while one holds the snake's attention the others leap at it and bite it. Although the snake attempts to retaliate it eventually succumbs to their bites, the final one usually delivered behind the head. Mongooses appear to have some resistance to snake venom. A Banded mongoose that had been bitten on the snout by a Snouted cobra showed few symptoms of envenomation two hours after the bite. However, some symptoms developed later but after treatment the mongoose showed no further ill effects.

There are other reptiles that prey on snakes, amongst them crocodiles. The diet of the Cape file snake consists largely of other snakes, including the more poisonous species. Dr Don Broadley recorded an instance of a Cape file snake having consumed a Herald snake, which in turn had eaten another snake. Cape and Snouted cobras are notorious for eating other snakes such as Puff adders. Some, like the shovel-snouted snakes, live largely on reptile eggs, including those of other snakes. Even the Giant bullfrog, *Pyxicephalus adspersus*, is known to eat small snakes, one having been recorded with 17 juvenile Rinkhals, *Hemachatus haemachatus*, in its stomach.

Only a small percentage of young snakes therefore actually reach maturity.

Growth can be fairly rapid provided food is plentiful and ambient temperatures stay between 21 °C and 32 °C. A Black mamba, which at hatching is approximately 45 cm in length, can attain a length of 1,5 to 1,8 m in one year, but normally growth is slower as cooler temperatures in winter retard development. Snakes continue to grow throughout their life but growth may slow down after the age of three to four years, and the snake may increase in girth.

Some snakes are relatively long lived in captivity and the larger species may attain ages in excess of 20 years. Cobras have been recorded living up to 15 years. It is doubtful that such longevity exists in the wild. It may have been possible before current levels of human population densities, but today a snake probably rarely attains an age of more than 10 years.

MYTHS

Snakes have and always will be the subject of legends. Very large and highly poisonous species are particularly favoured. However, even a harmless species such as the Cape file snake, *Mehelya capensis*, is considered deadly poisonous by some people. In Zimbabwe, where this snake is known by the Shona as 'Njoka ndala', it is said that its very presence near a kraal is an indication that someone there will die.

Pythons and anacondas were reported to attain incredible sizes by early explorers, with anacondas of 50 feet (16 m) having been recorded along the Amazon River. There is always speculation about which is the largest snake. Anacondas reach a length of 10 m while the Reticulated python from south-east Asia has been recorded growing slightly longer. Both are constrictors and the crushing power of these giant snakes is enormous but strangely enough few, if any, bones are broken, the prey dying from asphyxiation and cardiac arrest. A python need not anchor its tail in order to constrict as is popularly believed but uses its muscles only to achieve the desired effect. The two projections on either side of the body close to the tail are remnants of the former pelvic girdle and have nothing to do with the anchoring of the snake when it constricts its prey.

The Puff adder is said to strike backwards, but this is impossible as the fangs are mounted in the front of the mouth and are slightly recurved. It is therefore able to strike accurately in front and to either side.

Another popular tale is that cobras are capable of milking cows by twining themselves around the hind leg of the unfortunate animal and suck milk from its udder. This is also completely untrue as the snake is physically incapable of such suction, quite apart from being unable to climb up the cow's leg, even should the animal stand still for this to happen.

Throughout the southern half of Africa there are tales of a legendary crested snake that lives on rocky hills and either whistles, crows like a cock or bleats like a goat. It is said to be very large and to feed on dassies. This snake is imbued with extraordinary powers and beware the mortal that dares approach its lair. In Zimbabwe it is known as 'Murovhambhira' or dassie killer, a name also given to the Black mamba, *Dendroaspis polylepis*, while further north, Sweeney records the name 'Njoka tambala' from Malawi, and Livingstone in his book *Missionary Travels and Researches in South Africa* mentions a serpent called 'Noga-Putsane', which utters a cry like the bleating of a kid. In Zululand it is called 'inDhlondhlo'. The origins of these beliefs are obscure and there is no scientific verification of them despite numerous attempts to find this legendary creature, which is held in such awe by the local inhabitants.

Another interesting belief, encountered in Zimbabwe, concerns the Vine snake, *Thelotornis capensis*, which is called 'Kankamuti' or 'Kalikuni' basically meaning imitation wood. The very habits of this species in lying outstretched, often with more than a third of its body away from the supporting branch, have brought about the belief that the snake lies in wait for some unsuspecting person to walk by, whereupon it launches itself from the branch and arrow-like impales the body, killing the person.

The Red-lipped or Herald snake, *Crotaphopeltis hotamboeia*, is said to dig a hole in the ground and line it with bits of stone and pottery collected from the graves of people.

Throughout Africa snake myths and legends have been propagated by both black and white, the latter frequently reciting those of the former, usually embellished. It is strange that despite living with these animals people are unable to distinguish between harmful and harmless snakes, and have imbued them with such legendary powers. Such myths are not restricted to Africa alone but occur worldwide, also in China where the Chinese drink the blood of snakes to cure illnesses and use parts of the body for various remedies.

The explorer Sir Harry Johnston recorded in 1902 that African hunters used snakes such as the mamba and Puff adder to aid them in killing wild ungulates such as buffalo. The snake was kept stationary in the middle of a game track by nailing it through the tail to the ground, the enraged animal striking at passing animals. As many as 10 buffalo were said to have been killed by a single Puff adder in a day. The first body was discarded as being poisonous, but all subsequent ones were considered good eating. Although improbable, It is a pity that there is not more concrete evidence of this practice, which would have required careful and hazardous handling.

SNAKE CHARMERS

Perhaps further mention should be made of snake charmers, as this is a controversial subject, although most herpetologists agree that there is nothing mysterious about their activity. All snake charmers use live snakes, which are freely handled and in many instances made to perform by allegedly responding to music produced by a flute. Prior to such shows, poisonous snakes are subjected to the removal of the poison fangs and the reserve fangs, which lie behind the functional ones. These reserves move up the jaw and replace those lost or broken during prey capture. Such removal, if done carefully, will allow the animal to live for several weeks afterwards, but generally such operations result in death from septicaemia.

Another method of rendering a poisonous snake harmless is by sewing its mouth closed, but this method is not as popular as the former, although it does permit the snake charmer to put the snake's head in his mouth or wrap it around his head. Only in rare instances will snake charmers use poisonous species that have not been rendered harmless by these methods. In such instances the reptiles will have been in captivity for a long time and do not react as readily as newly captured individuals. There is always an element of risk, but since poisonous snakes often strike with a closed mouth and can become accustomed to being handled, if this is done slowly, gently and consistently, the risk can be minimised, particularly if the animal is already lethargic due to underfeeding and bad housing conditions.

Many snake charmers use harmless snakes such as egg-eaters and house snakes, which they put in their mouths and handle freely.

THE BLIND SNAKE FAMILY
TYPHLOPIDAE

The Typhlopidae or blind snakes are almost cosmopolitan in distribution, and contrary to most snakes have both dorsal and ventral surfaces covered by small scales. The only large scales present being those of the head. The eyes are vestigial and probably have limited functions such as being able to distinguish between light and dark. The mouth is small with peg-like conical teeth situated only in the upper jaw. There are about 160 species in six genera of which three genera and six species are found in South Africa. They are interesting animals ideally suited to their subterranean existence. The tail is very short with a pointed scale at the tip. The short tail enables the snake to reverse inside its burrow or is used as an anchor against which to push to promote sufficient force when making a new burrow. The snout is covered by a large scale, which in the genus *Rhinotyphlops* is depressed, and used to push through the soil. Coupled with a strong, stout cylindrical body these snakes are able to force their way through loose soil.

They range in size from a total length of 25 cm to the giant Schlegel's blind snake, *Rhinotyphlops schlegeli*, which attains a length of 90 cm and a girth to match. These snakes are mainly found under stones, logs and in mulch, but during the rainy season are often encountered moving about on the surface around homesteads or crossing roads. They are completely harmless and feed on invertebrates like termites, including their eggs, and worms. Like that of most burrowing snakes, the tail is very short enabling the animal to reverse inside its burrow with ease.

Little is known about the life of these snakes. All species appear to be either egg-laying or ovoviviparous. According to Sweeney in *Snakes of Nyasaland*, Schlegel's blind snake is oviparous, the young hatching after an incubation period of two to three months and up to 50 eggs are laid. In other species such as Bibron's blind snake, *Typhlops bibronii*, the eggs are retained in the oviducts, incubating internally until development is almost complete and then laid. The young hatch after two to five days. At the time of laying the eggs have a very thin, almost membranous opaque shell which is soft and pliable.

An interesting observation was recorded by C.R.S. Pitman in *A guide to the snakes of Uganda* where he quotes two scientists who observed an association between blind snakes and Driver or Matabele ants, *Dorylus* spp. The snakes apparently live in the nests of these ants, and even moved with them to new nesting sites. On the other hand, Sweeney records Driver ants attacking blind snakes, whose only defence is a pungent smelling discharge from the cloacal glands.

TOP: *Schlegel's blind snake becomes obese with age.*
BOTTOM: *Bibron's blind snake is common in the east and north of South Africa.*

An interesting recent addition to South Africa is the so-called Flowerpot snake, *Rhamphotyphlops braminus*, a species which is now almost cosmopolitan in the tropics, due its propensity for 'hitch-hiking' in the soil of flower pots, hence its common name. This species is unique because it is parthenogenetic, that is it reproduces without the fertilisation of the egg and only females are produced. In South Africa it has been recorded from Durban and Cape Town, but has been recorded more widespread along the east African coast.

THE THREAD SNAKE FAMILY
LEPTOTYPHLOPIDAE

Thread snakes or Leptyphlopidae are the smallest snakes, ranging in length from 15 to 25 cm and in girth from 1 to 3 mm. They weigh as little as 0,15 g. These little snakes are less specialised for a burrowing existence than the blind snakes and consequently have longer tails in relation to head and body. They are mostly found under stones and logs and among the roots of grass tussocks, and are frequently found in association with certain subterranean termite nests. These insects together with ants and their larvae appear to be their primary source of food. Thread snakes have blunt heads with very small jaws which only open wide enough to enable them to feed on such small creatures. In contrast to the blind snakes, they only have teeth in the lower jaw. The body is covered with small, uniform rounded scales and the eyes are vestigial and covered by scales, presumably only being able to discern between light and dark. Vestiges of the pelvis and femur remain as opposed to the blind snakes where only remnants of the pelvis are found.

These snakes are primitive, and have a wide distribution from the Americas to Africa. Only two genera are recognised with some 90 species, of which one genus and nine species having been recorded from South Africa, three of them endemic. Although most are dark in colour, brownish black to black, the scales of some have a pale margin, giving them a chequered appearance. This becomes more pronounced after being exposed on the surface for a while, so that moisture trapped between the scales evaporates. Two species are brownish pink, one of which, *Leptotyphlops longicauda*, has a particularly long tail. Sexual dimorphism is apparent with females having a shorter tail than males.

The Long-tailed thread snake, Leptotyphlops longicauda, *is mostly found under stones and rotting logs.*

They are oviparous laying up to seven elongated eggs, which in some species are attached to one another by thin strands of membrane, like a string of sausages. These are usually laid under stones, in moribund termite mounds and logs. Hatchlings are usually found between January and April.

THE PYTHON AND BOA FAMILY
BOIDAE

Of great fascination to most people are the giants among snakes, the boa constrictors, pythons and anacondas. Although they are the largest of our snakes, they are surpassed by some of their extinct ancestors, such as *Gigantophis*, which may have reached between 13 and 16 m in length. There is only one family of these primitive snakes, which came into existence some 100 million years ago. The family is divided into two subfamilies, namely the boas and the true pythons, the former are viviparous and the latter oviparous. There are about 23 genera and about 70 species of boas and pythons distributed throughout the tropics both north and south of the equator and in some instances even in the temperate zones.

The boas are mostly confined to the New World except for the Sand boas of North and East Africa and three species in Madagascar.

In South Africa only one species of the genus *Python*, the Southern African python, *P. natalensis*, is found, while another species, the Angolan or Anchieta's dwarf python, *P. anchietae*, also occurs in Namibia. Two other species occur further north including the African python, *P. sebae*, and the Royal or Ball python, *P. regius*, but the genus is widespread across south-east Asia.

The family Boidae is generally considered as primitive as they have vestiges of the pelvic girdle and the hind limbs appear in the form of spurs on either side of the body anterior to the cloaca. In addition, the dorsal scales are small and uniform, while only the midventral scales are enlarged differing noticeably from their neighbours, but have not yet reached the size found in more recent families. The family includes a great variety of forms and habits, including arboreal, burrowing, terrestrial, rock living and even semi-aquatic species.

As mentioned, only the Southern African python is found in South Africa. It vies for third place as one of the world's largest snakes, along with the Australian Amethystine python, *Liasis amethystinus*, reaching an estimated 8 m in length. Only two other species are larger, namely the Reticulate python, *P. reticulatus*, and the Anaconda, both of which are known to reach 10 m. However, such lengths are regarded exceptional today. The largest Southern African python I have seen was in Zimbabwe, measuring 5,6 m. Generally adults average about 3 to 4 m in length and can be fairly heavy. A 4,3-m female had a mass of 44 kg. This, however, depends on the time of the year and the condition of the animal, because the same python weighed 10 kg less at the end of the following winter.

Pythons may grow fairly rapidly during the first few years of life and a hatchling, which measures between 45 to 60 cm in length, could reach 1,5 m in length in one year. This rate does slow down later and a 3-m snake may be three years old or more, depending on the food supply. Most natural mortality occurs after the young hatch and begin to fend for themselves. The Southern African python inhabits a variety of habitats from grasslands to bushveld but shows a preference for rocky hills and areas close to water. It lives in holes in the ground, such as antbear and springhare burrows, while crevices between and under rocks are frequent refuges. They will also lie up in dense vegetation.

The python is mainly diurnal and is often found lying in front of its hole basking in the sun. However, when temperatures rise too high it retires to cool down in the burrow. Those living close to permanent water can frequently be seen lying submerged with only the head or tip of the snout protruding. They take readily to water if disturbed and swim well, and if surprised will dive down to the bottom and remain there for as long as an hour without coming up for air.

The python is as such a terrestrial animal, but climbs trees with ease. This applies especially to younger snakes up to 3 m in length. On the ground they move by means of the belly scales in rectilinear fashion making it easy to follow their spoor in sandy areas as it runs in a straight line.

Pythons feed on a wide variety of prey depending on their size. Younger snakes feed mainly on mice, birds and frogs whereas the adults consume larger prey. This varies

The Southern African python is the largest snake in South Africa. It is widely distributed in the northern and eastern parts of South Africa, and was relatively recently reintroduced into the Eastern Cape.

The head of a Southern African python, showing the heat-sensitive pits between the scales at the front of the snout.

from cane rats, dassies, porcupines and hedgehogs to larger animals such as impala. There are even accounts of leopards being attacked. Birds also form an important part of the diet and guinea fowl and Egyptian geese are regularly captured and even monitor lizards are eaten.

Normally prey is caught while the snake lies in wait, curled up under a bush near a path or among reeds and long grass or even when lying partially submerged in the water. To assist in prey location, which is a function of sight and scent, the python has additional sensors located in pits at the tip of the snout. These are heat sensitive and advise a snake of the presence of a warm-blooded animal nearby. The prey is caught by a sudden strike and with a mouth that can open to an angle of 130°. It bites the prey, holding on by means of the sharp recurved teeth. The power behind the strike may force the prey off balance and this gives the snake a chance to throw two or three coils around it, suffocating it by increasing pressure, the prey dying from asphyxiation and cardiac arrest. Once the animal is dead, the snake relaxes its hold and begins to nose it until satisfied as to the position of the head and then

TOP: *The wide-open mouth of this striking python gives an indication of the efficacy of the bite.*

BOTTOM: *The skull of a python showing the recurved teeth adapted to provide maximum hold on its prey.*

A python hatchling in the process of emerging from the egg.

swallows it head first. The elastic ligament binding the two lower halves of the jaw expands to a great degree, enabling the snake to swallow astonishingly large prey. An abundant supply of saliva assists this process. The rate of swallowing is dependent on the size of the prey relative to the size of the head, but can take 30 minutes or longer.

Once the python has ingested a large meal it will move to a secluded, shady spot where digestion begins. This can take up to 10 days if the prey is large. Due to the highly acidic stomach juices even bones will dissolve; only hair, horns and hooves are not digested and can be found as remnants in the faeces. Occasionally reports are received of pythons with the horns of a buck protruding through the skin. These will eventually work their way out and the holes will heal. During the warm months of the year the python will eat as regularly as possible, in order to build up a sufficient supply of body fat to tide it over the winter months, particularly in South Africa where winter temperatures can be low.

Pythons can live for long periods without food and frequently fast in captivity. There is an instance recorded where a Southern African python refused food for a period of two years and nine months, but this is really exceptional. Normally a period of four months during winter is the rule, but up to one year is not unusual. When hibernating, pythons do not drink, which they readily do during summer.

It appears at present that pythons do not have a specific reproductive season. There are cases of females laying their eggs in January, March, April, June, October and November, though there may be a possible peak during March and April.

The eggs are laid in a variety of sites, such as hollows under trees, thick bush, dense grass and reed-beds as well as antbear and springhare burrows. Between 20 and 100 eggs may be laid, although a clutch of 20 to 30 eggs is considered normal. The eggs are round and the shell is strong, but pliable. They measure approximately 112 mm in diameter and have a mass of 265 g.

Pythons are the only snakes that actually incubate their eggs, the female being able to raise her body temperature to slightly higher than ambient temperature. She lies coiled about the eggs and stays in this position for six to eight weeks, only seldom leaving them. The embryos in the eggs develop considerably prior to laying and have an average length of 49 cm and a mass of 72 g when the eggs are actually laid. A large number of the eggs may be infertile and up to 50 per cent of the clutch may be lost. After an incubation period of six to eight weeks, before the young are ready to hatch, the female leaves the eggs. The hatchlings, armed with a minute egg tooth on the tip of the snout, cut through the shell and then rest with the head protruding from the egg. If disturbed they may retreat into the shell and only emerge when all is quiet. The hatchlings are between 45 and 60 cm long and generally remain together for a number of days, sometimes in the company of the female, who may still inhabit the same hole. With the exception of the incubation of the eggs, there is no parental care.

The snake families discussed so far are regarded as primitive because they all show vestiges of the pelvic girdles as well as of the hind limbs. The following families lack such vestiges and are considered more advanced as a result. They include all of the harmless species as well as poisonous snakes. Among the most unique is the family Atractaspididae, which consists mostly of species adapted to a burrowing lifestyle.

THE BURROWING SNAKE FAMILY
ATRACTASPIDIDAE

With the exception of the centipede-eating snakes of the genus *Aparallactus*, most of the snakes of this family have modified snouts as an adaptation to a burrowing existence. These include the Burrowing asps of the genus *Atractaspis*, the purple-glossed snakes and related species of the genus *Amblyodipsas* and the quill-snouted snakes of the genus *Xenocalamus*.

There are three species of purple-glossed snakes in South Africa of which the Common purple-glossed snake, *Amblyodipsas polylepis*, is the most widespread and frequently seen species. The specific name means 'many scaled' and as its common name suggests it is uniform glossy black with a purplish sheen, both dorsally and ventrally, which becomes an opaque bluish-grey just prior to shedding its skin. The tail, like that of other burrowing snakes, is short and typically blunt. The head is depressed dorsally and shovel-shaped. They are normally not seen above ground except during the rainy season, when following the passage of rain they may emerge to move about and forage on the surface. They are comparatively sluggish and if molested will move about jerkily. However, if desperate they may hide the head under the body coils and raise the stumpy tail, waving it about in the air, attempting to draw attention to this less vulnerable part of its anatomy. Their main food consists of burrowing snakes such as blind and thread snakes, worm lizards and burrowing lizards.

The quill-snouted snakes are limited to two species with three subspecies in South Africa. The name quill-snout is derived from the flattened, elongated and sharply pointed head, while the body is cylindrical and has a short tail. The most common of the two quill-snouted species is the Bicoloured quill-snouted snake, *Xenocalamus bicolor*, which is comprised of three subspecies, *X. b. bicolor*, *X. b. lineatus* and *X. b. australis*, of which the latter is endemic to South Africa and is restricted to the Waterberg and environs in Limpopo Province. The typical race occurs in Namibia and Botswana, just entering South Africa in the north-west, the race *X. b. lineatus* is chiefly found in Mozambique, the lowveld and Zululand. These snakes are usually black above and yellow below, with the exception of the Transvaal quill-snouted snake, *X. transvaalensis*, which has the upper body scales black edged with yellow, giving a chequered appearance. This snake is rarely seen although it is distributed from central Mozambique and south-eastern Zimbabwe to north-eastern Limpopo Province and further south to Zululand. Its scarcity can in part be attributed to its

The Purple-glossed snake showing the stumpy tail which it may raise and wave about while hiding its head amongst its coils in the hope of confusing a predator.

The narrow pointed head of the Bicoloured quill-snouted snake, an adaptation to a burrowing existence.

fossorial or burrowing habits. It feeds mostly on amphisbaenians. These snakes are inoffensive and never attempt to bite.

TOP AND CENTRE: *The arched neck of Bibron's burrowing asp indicates its readiness to defend itself. Centre photo: J. Marais*
BOTTOM: *The Common centipede-eater is widespread in most mesic environments.*

Until recently the burrowing asps, genus *Atractaspis* were incorporated into various families such as the Viperidae and Colubridae, but research into the properties of the venom as well as their genetic makeup shows that this is incorrect. They are now recognised as being intermediate between the quill-snouts and the centipede-eaters, all belonging to the same family. Two species occur in South Africa, of which Duerden's burrowing asp, *Atractaspis duerdeni*, is exceptionally rare and no details of its habits and behaviour are known. Bibron's burrowing asp, *A. bibronii*, is the most common species encountered and is characterised by having a shovel-shaped head and short tail mostly terminating in a spike-like scale at the tip. The body is long and slender although the snake does not exceed 90 cm in length. The eyes are small but still functional. This snake has an elaborate venom apparatus in which the poison fangs are very long and therefore lie pointing backwards, along the upper jaw, and just below the eye.

The toxic venom is injected in a unique way, by a twisting rearward stabbing action. Due to the length of the fangs only one fang is imbedded, as the lower jaw is twisted to one side. It is therefore impossible to hold this snake behind the head and many people have regretted their encounter with this snake. The bite is particularly unpleasant and painful, but does not appear to be lethal. If picked up by the tail, the hardened scale tip may be jabbed into one's finger causing one to drop it in surprise. In colour these snakes are mostly black above and white below, although completely black individuals are also known. It feeds mainly on thread snakes and is found under stones, in rotting termitaria and under rotting logs. These snakes are also oviparous.

Two unique members of this family occurring in South Africa are the centipede-eaters, of which the Cape centipede-eater, *Aparallactus capensis*, is the most common and widespread. This small mostly less than 60 cm long snake is particularly common in decaying termitaria and as many as five have been found in a termite mound. They also live in and under rotting logs and under stones and other debris. These small snakes feed mainly on centipedes, which are caught by biting them behind the head and the poison is worked in, in the typical manner of back-fanged snakes. The snake may also chew backwards and forwards along the length of the centipede. The poison takes effect rapidly and once the prey is immobilised it is swallowed. Centipedes frequently occur in termitaria, which may account for the large number of these snakes found in them. Despite this, such decaying termite mounds are very important refuges for many species of snake on the highveld during the winter months, protecting them from the cold and from veld fires, which occur frequently at this time. One species of dwarf shrew, *Suncus varilla*, is mostly recorded from such mounds, which are probably essential for its survival on the highveld.

ADVANCED SNAKES

The great majority of living snakes belong in this category and more than two-thirds can be placed in a single family, Colubridae.

THE ADVANCED SNAKE FAMILY
COLUBRIDAE

There are about 300 genera and some 1 500 species in the family distributed worldwide, only being restricted by the line of permafrost, which precludes hibernation underground. Of these some 22 genera and 53 species occur in South Africa of which nine species are endemic or near endemic. As is the case with all of the remaining families, these snakes are covered by fairly large overlapping scales dorsally, but ventrally these are transversely much enlarged to form a single row. There are nine head shields symmetrically arranged and rows of backward curved teeth on the maxilla, palatine and pterygoid, as well as along the dentary bones.

The family has been variously split into subfamilies according to various interpretations of specific characteristics. This account follows that adopted by Bill Branch in a *Field guide to the snakes and other reptiles of southern Africa* incorporating most species into four subfamilies, excluding three genera whose affinities are not yet clearly established. These subfamilies include the Lamprophinae, Natricinae, Psammophinae and Colubrinae. It is uncertain at this stage where the genera *Prosymna*, *Montaspis* and *Amplorhinus* fit in, but the latter two are endemic or near endemic to South Africa.

Sundevall's shovel snout preys on the eggs of other reptiles which may be swallowed whole.

A subspecies of Sundevall's shovel-snout, Prosymna sundevalli lineata *mostly found in the lowveld of Limpopo and Mpumalanga provinces.*

The snakes of this family are therefore extremely diverse in form and habits, characteristics that tend to be related to the habitats they occupy. There are burrowing, terrestrial, rupicolous, arboreal and aquatic species. Among the most interesting are the small, cylindrical, short-tailed shovel-snouts of the genus *Prosymna* of which five species are known to occur in South Africa. They occupy a great range of habitats from semi-desert to coastal forest, but with the possible exception of the Mozambique shovel-snout, *Prosymna jani*, of which nine individuals per kilometre were recorded in coastal forest, are nowhere common. These are burrowing snakes with depressed heads and small eyes, which feed on the eggs of other reptiles although some species also prey on lizards. Sundevall's shovel-snout, *Prosymna sundevalli*, is among the most widespread with two subspecies, the one occurring from the Cape to the bushveld of the Limpopo Province while the other is mostly found from Maputaland northwards to Zimbabwe. These snakes are totally defenceless and as a result have adopted a peculiar method of defence when molested, by coiling and uncoiling violently thereby deterring possible predators. They are egg-laying but very little else is known of their habits.

Another very interesting species is the Cape file snake, *Mehelya capensis*, so-called because the body is triangular in shape, with a pronounced row of enlarged scales along the vertebrae. This completely inoffensive snake is nocturnal and feeds preferably on other snakes but also includes toads and mice in its diet. It kills snakes by biting them along the length of the body until they are immobilised and then swallows them whole. It feeds on all species, even venomous ones and seems to be immune to their venom. This grey snake with a narrow vertebral stripe may reach a length of 1,5 m and has little defence against predators. If molested it will hide its head under its coils and produce a grey, pungent and viscous fluid from its cloacal glands.

It is a rare species occurring in very low densities throughout its distribution. In the Nylsvley Nature Reserve in Limpopo Province only four individuals were found in three years. It is unfortunate that this completely harmless snake is considered to be poisonous by the indigenous people as well as Europeans. Its distinctive shape and slow movements should enable everybody to recognise and conserve it. There is a second species, the Black file snake, *M. nyassae*, which also retains the triangular body form but is much smaller only reaching 60 cm in length. It is a shiny black with skin interstices between the scales a dark pink or purplish-pink. The species appears to be more common than the former and feeds mostly on lizards.

There are two species of slug-eaters, which as the name implies feed largely on slugs and snails and can in fact extract the latter from its shell. They have a small head and a stout body. It is amusing to observe these small snakes, which reach a length of 37 cm, blowing bubbles after feeding on snails. The Common slug-eater, *Duberria lutrix*, is the most widespread species and

TOP: *The file snakes are a unique genus of snakes, of which one of the most striking and rare is the Cape file snake.*
BOTTOM: *An unusual defensive posture of the Common slug-eater, hiding its head in its coils.* Photo: Richard Boycott

The head of a Spotted bush snake, one of the most widespread of the green snakes.

occurs mainly along the moist eastern side of South Africa, where the rainfall is high and food is plentiful. In parts of their range such as Sabie in Mpumalanga and George in the southern Cape, they are quite abundant, usually taking refuge under stones, logs or vegetable debris. In contrast the Variegated slug eater, *D. variegata*, is much more restricted in range being limited to coastal forest in Zululand and Maputaland.

In contrast to most snakes in this family, these little snakes are viviparous and may produce as many as 20 young at a time. They are sluggish, nocturnal in habits and completely harmless with no defence system. Like the shovel-snouts this snake also coils and uncoils violently hoping to scare off a potential predator.

The Green snakes are among some of the most beautiful of our snakes. They are bright green to blue-green in colour and have a relatively long tail. They are normally found close to water with the exception of the Spotted bush snake, *Philothamnus semivariegatus*, which is usually found away from water. This species is well adapted for climbing trees, having keeled ventral scales, which assist the snake in maintaining a foothold. This adaptation is shared with the Natal green snake, *P. natalensis*, which tends to occur on small shrubs and bushes flanking streams. Only the Green water snake, *P. hoplogaster*, is mostly terrestrial and associated with water bodies, although it is often found some distance away from water. The green snakes swim and dive well, feeding mostly on frogs. The Spotted bush snake,

Green water snakes are often confused with the Boomslang and Green mamba.

The Black water snake, an uncommon species along the escarpment of KwaZulu-Natal and Mpumalanga.

The Brown water snake, a widespread species in South Africa from the coast to the highveld.

although also able to swim and feeding mostly on frogs, tends to have a wider diet including fledgling birds as well as lizards. It may become overexcited when following prey and on one occasion an individual of about 75 cm in length was found hanging from the beam of a garage roof with a large Grey tree frog, *Chiromantis xerampelina*, in its jaws. As the frog was at least three times too large for the snake and consequently quite strong, it managed to struggle free and fell some 2 m to the ground. Without a moment's hesitation the snake launched itself after the frog, landing with a thud and gave chase, but the frog, suffering only a few bleeding teeth marks, managed to escape.

The Green snakes are frequently confused with the Boomslang, *Dispholidus typus*, and the Green mamba, *Dendroaspis angusticeps*, but they rarely reach more than 90 cm in length, with the exception of the Spotted bush snake which may reach 1,2 m. They can be distinguished from the Boomslang by their smooth shiny scales, while the Green mamba is restricted to suitable habitat along the coast in a narrow strip from Port St Johns northwards to Mozambique, not extending further inland than about 30 km from the coast. Elsewhere all smooth green snakes are harmless, although they do bite if molested. They are oviparous, laying from five to 12 eggs.

Three other species also usually associated with water include the Brown water snake, *Lycodonomorphus rufulus*, the Black water snake, *L. laevissimus*, and the White-lipped water snake, *L. obscuriventris*. The former is the most common and has a wide distribution in South Africa from the southern Cape to Limpopo Province with an outlier population in eastern Zimbabwe. Despite their name, they tend to be olive brown or olive above and pink to pinkish yellow or even yellow below. Like most snakes the males have longer tails than the females. These snakes occupy a wide range of habitats especially along rivers but also dams, pans and gravel pits containing water. A nocturnal snake, it lies up by day in or under any shelter it finds, usually stones, in holes, under debris or piles of rotting vegetation. It feeds only on frogs and is a voracious feeder consuming several in quick succession if the opportunity arises. It hunts in and out of the water and is reported to even catch its prey and swallow it under water, but mostly comes to the surface to do this. These snakes are egg-laying, the eggs being laid under stones in moist areas near streams. The young emerge after an incubation of about two months and measure 15 cm at hatching.

The other species are much more uncommon, the Black water snake occurring mostly along streams in KwaZulu-Natal and the highlands of the Mpumalanga escarpment, while the White-lipped water snake is largely restricted to pans and marshes in the lowveld of Mpumalanga and adjacent Swaziland, Mozambique and south-eastern Zimbabwe. Both of these species are also nocturnal, because their prey, consisting mainly of frogs, is also mostly active at night. The former also eats fish, which it catches and swallows under water. All of the water snakes are harmless although some may bite if molested.

Some of our rarest snakes are Fisk's house snake, *Lamprophis fiskii*, the Yellow-bellied house snake, *L. fuscus*, and Swazi house snake, *L. swazicus*, all of which are listed in the South African *Red Data Book* as rare. Little is known of the habits of these species. Fisk's house snake is a very attractive snake with black scales edged with yellow or yellow-orange. It is a small snake and is confined to three small areas in the Greater Karoo and Namaqualand, apparently living in burrows and holes underground, surfacing to hunt at night. The Yellow-bellied house snake occurs much more widely but in scattered localities from Cape Town to Mpumalanga mostly in grassland and fynbos living in moribund termitaria and sometimes under stones. It is mostly olive green above with a yellow belly.

TOP: *Fisk's house snake is one of three rare species of house snake found in South Africa, and probably the most attractive.*
BOTTOM: *The rare Swazi house snake, a species of rocky outcrops along the eastern escarpment from Swaziland to Limpopo Province. Photo: Richard Boycott*

The Swazi house snake is also very rare having only been recorded from the Drakensberg escarpment in Mpumalanga Province with an outlier population at Turfloop, near Polokwane in Limpopo Province. Originally described from north-eastern Swaziland in 1970, it is a slender snake with a head distinctly broader than the neck. The eyes are cat-like with a vertical pupil typical of many nocturnal snakes. It is mostly a uniform reddish brown colour above and paler below becoming yellow brown along the ventral scales. It occupies crevices between and under rocks on rocky outcrops mostly at altitudes between 1 400 and 1 900 m above sea level.

There are four other species of house snake in South Africa, one of which, the Brown house snake, *Lamprophis fuliginosus*, is the most common, almost occurring throughout the country and northwards to the Sudan and west to Cameroon. It is mostly varying shades of brown above and off-white to yellowish pink below and is easily recognisable by the thin white line extending from the tip of the snout to above and behind the eye. On average it is mostly small but may reach a length of 1,2 m on rare occasions. Such large specimens are rare today as many individuals are killed by misguided people before they attain such lengths. A nocturnal snake, it shelters by day under any cover but in particular under and in rotting logs, rotting antheaps, under stones, among building refuse as well as in holes. They feed mostly on lizards while young but as they increase in size they depend more on rodents and are therefore a boon to suburban dwellers as well as farmers. They appear to live commensally with humans and are likely to benefit due to the increased availability of prey and refuges. The prey is actively hunted, constricted and swallowed.

The Brown house snake is oviparous laying up to 16 eggs, but usually 6 to 10. The eggs are elongated and oval, averaging about 30 mm × 15 mm, and are mostly laid during October and November, although they may be laid at other times of the year. The number of eggs laid by a female is correlated to total length. The incubation period is variable according to environmental conditions but is mostly between eight and 12 weeks.

CLOCKWISE, FROM TOP LEFT:
The Aurora house snake, one of the more attractive of our snakes.
A Cape wolf snake, a nocturnal and widespread species.
Unfortunately the Mole snake's large size and truculence often leads to it being killed. It is completely harmless and an inveterate rodent catcher.
Rare in South Africa, the Variegated wolf snake occurs in the lowveld of Limpopo and Mpumalanga extending into northern KwaZulu-Natal.

Among the more beautiful of our snakes is the Aurora house snake, *L. aurora*, which is widespread in the eastern half of the country. This snake is olive green with an orange stripe down the middle of the back, most pronounced when young but becoming less striking with age. Generally a highveld species it is mostly restricted to the grasslands of Gauteng, Mpumalanga, Free State and the midlands of KwaZulu-Natal but also occurs in several separate populations in fynbos areas of the southern Cape. It is a completely harmless species feeding on lizards and mice.

Less well known is the Spotted house snake, *L. guttatus*, which is distributed from the Western Cape eastwards along the fold mountains to the Eastern Cape, KwaZulu-Natal and into Mpumalanga and Limpopo provinces as far north as the Soutpansberg at altitudes up to 2 300 m above sea level. Relict populations also occur in southern Namibia. This snake is quite variably coloured, from grey-brown to yellowish and pinkish brown above with several series of reddish to dark brown blotches or spots which are arranged in either alternating or confluent pairs, which may form a zigzag pattern down the back.

It is normally associated with rocky outcrops living by day in crevices between and under rocks. Its head is flattened as an adaptation to its rupicolous existence enabling it to squeeze into narrow spaces. One individual was recorded from the summit of the Madhlangampisi Mountains on the border of Mpumalanga and KwaZulu-Natal, inhabiting a crevice in the rocks at the edge of a high cliff. It feeds on similar prey to that of the other house snakes and is also oviparous but lays relatively few eggs ranging from three to six in number. On one occasion during a survey of the reptiles and amphibians of the former Transvaal, a skink was disturbed when a rock was lifted and it ran under another rock where it was promptly seized and eaten by a Spotted house snake which had taken refuge there.

A common and widespread snake is the Cape wolf snake, *Lycophidion capense*, so named for the elongated teeth typical of this genus. It has a very wide distribution throughout the eastern half of South Africa, but is absent from the arid western parts, its distribution extending north into East Africa. A small snake rarely reaching a length of 37 cm, it is black above, most frequently with a small white spot on each scale. The scales on the head are marked with a filigree pattern. The head is depressed and the snake has a short tail. They feed solely on lizards including both nocturnal and diurnal species. Usually found under stones and in rotting antheaps, they are completely inoffensive, but if molested may move jerkily, and can therefore be confused with Bibron's burrowing asp which is a poisonous species.

A much rarer species is the Variegated wolf snake, *L. variegatum*, which is listed as rare in the *Red Data Book* and occurs from Zululand northwards through Swaziland to the lowveld of Mpumalanga and the bushveld of Limpopo Province north to Zimbabwe and southern Zambia. It is usually much more distinctly marked with white than the previous species. It appears to be more secretive and perhaps restricted to rocky outcrops, feeding mostly on the same prey as the Cape wolf snake.

An impressive snake belonging to this family is the Mole snake, *Pseudaspis cana*, which is distributed almost throughout southern Africa extending as far north as Angola in the west and Kenya in the east. It has a large heavy body and a small, slightly depressed head, and varies in size and colour throughout its range, but is largest in the Western and Northern Cape provinces where individuals of 2 m, usually shiny black, may be found. Further north, black specimens may be found in the south-western parts of North West Province becoming varying shades of brown, often blotched and marked, elsewhere in that province, Gauteng and Mpumalanga. In the northern Limpopo Province grey coloured specimens may be found. Juveniles are variegated and blotched and do not resemble the adults, but the markings disappear with maturity.

It takes refuge in holes in the ground, usually those of other animals such as mice, ground squirrels and springhare burrows. It feeds mostly on burrowing mammals such as mole rats, gerbils, moles and mice, while the young may eat lizards and occasionally reptile eggs. The adults may sometimes be seen lying with the front part of the body down a mole rat burrow, waiting patiently for the animal to come and close up the hole. This rodent has four well-developed incisors, which

TOP: *A grey form of the Mole snake from the northern Limpopo Province.*
BOTTOM: *Juvenile Mole snakes do not resemble the adults, only gradually losing the blotched and speckled appearance as they grow older.*

could inflict serious injury if the snake is not careful. These snakes are excellent rodent catchers, but if molested can be quite aggressive and bite freely, however they are not poisonous, only drawing blood in the event of a bite. They are viviparous and produce between 30 and 50 young although a record 95 has been reported. The young are born during the summer months and may be 200 mm in total length. This snake should be protected throughout its range.

A unique and very interesting group of snakes are the egg-eaters of which three species occur in South Africa. As their name suggests these snakes, which include the Common egg-eater, *Dasypeltis scabra*, Brown egg-eater, *D. inornatus*, and the East African egg-eater, *D. medici*, feed exclusively on birds' eggs. They are mostly slender, small to medium-sized snakes ranging from 30 to 120 cm in length. The pupil is vertical and cat-like. The best known of these snakes, the Common egg-eater varies considerably in colour from one area to another, but usually has rhomboidal markings along the back. Immediately behind the head is a V-shaped forward directed marking. This mark is also shared with the night adders but the egg-eater is more slender and elongated and the dorsal scales are keeled, whereas the night adder is relatively stout and has a velvety sheen to its scales. The mimicry is carried even further because the egg-eater when disturbed will coil and hiss loudly, flatten its head and open its wide mouth, showing the black interior. At the same time it makes clumsy lunges in the direction of the aggressor. This is all a sham because the egg-eater only has a few rudimentary teeth and the hissing sound is produced by the opposing action of the body scales rubbing against each other as the body is continually in motion. The whole appearance and aggressiveness is therefore very much like that of an adder.

Adults feed almost exclusively on birds' eggs – an individual of 90 cm in length may even manage fowl eggs, but reptile eggs are also taken. These reptiles are well adapted to this diet lacking all but a few tiny rudimentary teeth in the centre of the roof of the mouth. The jaws are capable of stretching exceptionally wide due to the elasticity of the ligaments and skin of the gular and neck. The first five or six vertebrae have special downward projecting processes. When the snake finds an egg it flicks its tongue along it to assess whether the egg is fresh, as rotten eggs are not eaten. If fresh, the snake will open its mouth and gradually work it around the egg, so that the whole egg is engulfed. As the egg is worked down the throat by muscular action it comes into contact with the bony projections. Muscular contractions press it against the serrated edges, breaking the shell. Continual muscular contractions crumple the egg and the contents are swallowed, while the broken shell is compressed and ejected through the mouth, likewise by muscular action. A valve in the oesophagus prevents the return of the liquid. Soft-shelled eggs of snakes and lizards are apparently swallowed whole. These habits are intriguing, but even more so is how does an egg eater find sufficient eggs to keep it alive and see it through winter? In certain habitats on the highveld Common egg-eaters are particularly common in decaying termitaria and as many as 10 may be found in an area of less than a hectare. How do these snakes find food in such a small area? It is likely that these snakes come to these sites from the surrounding areas but it still means that there must be a sufficient abundance of nesting birds during summer to sustain these snakes. The nesting period of birds on the highveld coincides with spring and summer and the snakes must therefore find sufficient nests at this time to enable it to build up fat

The Common egg-eater has no defence except mimicry of the Common night adder and an aggressive display including defensive strikes displaying the dark interior of the mouth.

reserves to tide it over the winter period from mid-April to mid-September or even later. The snakes appear to have a very keen sense of smell as a nest with eggs laid that day may be predated the same night.

Egg-eaters are egg-laying, producing from five to 25 eggs during summer, the young hatching some two to three months later and measuring 220 to 260 mm in length. Although they are sustained by remnants of the egg yolk for a while after hatching, once this has been utilised the hatchlings must find suitably sized eggs to tide them over the winter months.

In the past those snakes with modified teeth in the form of a primitive venom-conducting apparatus, including modifications to the salivary gland for the production of venom, were all placed together in one

One of the most widespread South African snakes, the Herald snake is also innocuous despite an aggressive display when molested.

A Herald snake eating a frog. Note the distended skin of the throat and the use of the fang to assist the ingestion of the prey. Photo: Richard Boycott

subfamily the Boiginae or back-fanged snakes. It is now apparent that such modifications appeared more than once, independent of relationships. The following snakes all have such a modified tooth structure, with the fang found either just below or just behind the eyes. From one to three teeth may be grooved along the front, along which the venom is conducted.

Probably the most common and well known of these back-fanged snakes is the Herald or Red-lipped snake, *Crotaphopeltis hotamboeia*. The name Herald is derived from the newspaper *The Eastern Cape Herald*, which first reported on the existence of this snake. Its other common name refers to the colour of the upper lip, which may be red or orange-red in parts of its range. However, these colours occur only in some parts of South Africa such as Gauteng and KwaZulu-Natal, but fade northwards and to the south, becoming white. It is not a large snake, mostly 40 to 70 cm long, but reaching a length of up to a metre in exceptionally large specimens.

This snake is mistakenly believed to be extremely venomous in South Africa due to its habit of flattening its head, showing the shiny black sides of the head, drawing it back into a typical menacing adder-like pose. At the same time the lips are flared to present an imposing stance to a potential threat. In this position the snake strikes freely at the attacker, often without biting. This is, however, just a pose as its venom is ineffective on humans and any other predator such as a mongoose or genet, and is sufficient only to kill its prey, such as toads. Like that of most other back-fanged snakes, the venom needs to be chewed in, in order to be effective.

The tiger snakes are amongst the most attractive species, two species being present in South Africa, both similar in colour and with wide distributions, the Eastern tiger snake, *Telescopus s. semiannulatus*, extending from Mozambique, Botswana and eastern Namibia south to the North West, Northern Cape and KwaZulu-Natal provinces, with a subspecies *T. s. polystictus* in Namibia extending south to the Northern Cape. The second species is Beetz's tiger snake, *T. beetzii*, named after a German geologist who worked in Namibia from 1914 to the 1930s. It occurs from southern Namibia south to the Northern Cape and across to the south-western Free State. The body of these snakes is brownish orange to orange with narrow to broad black bars evenly spaced down the back, the numbers of bands differing between the species. The head tends to be the colour of the body, and is broad and slightly flattened. They have large cat-like eyes with vertical pupils, typical of nocturnal snakes. These snakes are mainly terrestrial or partly arboreal while Beetz's tiger snake is largely rupicolous or rock living.

Slow moving, they can become fairly aggressive if molested, frequently biting without any warning. The Eastern tiger snake is the largest, reaching a length of a metre under exceptional circumstances. They feed mostly on lizards, as well as small birds and mice and have been recorded taking small bats under the eaves of huts.

Two back-fanged snakes occurring in South Africa, which rank among the more poisonous snakes, are the Boomslang, *Dispholidus typus*, and the Vine snake,

Different species and subspecies of tiger snakes in southern Africa are mostly similar in colour, differing amongst others in the number of crossbars down the back. This is the Eastern tiger snake, a species with a wide distribution in the northern parts of the country.

Thelotornis capensis. Both species are widespread in South Africa, especially the former, which is found virtually over the whole of the Republic with the exception of the Karoo and south-eastern Free State. It is a large snake attaining a possible length of 1,8 m but is mostly 1,2 to 1,4 m in length. It is variable in colour from plain greyish-black, to varying shades of brown and green. In certain areas such as the eastern and southern Cape, spotted or speckled forms occur where each scale is edged with a yellow spot. The dorsal scales are keeled and characteristic of this snake, as is the relatively large rounded head and large eyes, serving to differentiate this species from other green-coloured snakes. Sexual dichromatism is infrequent in snakes but is manifest in the Boomslang with females always brown, while males may be any other colour such as greyish-black but are most often green. Juveniles have large rounded heads with green eyes, the heads being dark brown above, which peters out posteriorly.

The Boomslang is a tree snake but is equally at home on the ground and as a result is often killed on the roads. It is an agile fast-moving snake, especially in the trees and feeds mainly on chameleons, but birds, eggs and fledglings are also eaten. It is quite common to find a Boomslang among a colony of weaverbird nests with the head inside a nest. On one occasion, while camping under such a colony which had attached their nests to the fronds of an iLala palm at Evangelina in north-western Limpopo Province, we were surprised by a large green Boomslang that landed with a great crash among the dishes on the breakfast table. It had obviously attempted to get at one of the nests and being harassed by the parent birds and helpers had lost its purchase on the frond and slipped off.

These snakes are not aggressive and very reluctant to bite, usually attempting to flee as quickly as possible upon being disturbed. If molested, they puff out the skin of the throat and with the head held up gradually ease away sideways until at a safe distance when they lower the head to the ground and dash off. Although the gape of the mouth enables the snake to bite at any part of the body, it usually has to get its fangs in position before any venom can be injected. It therefore has to chew the

The Boomslang is a notorious species, largely unfounded, as it is a very shy and retiring snake. Females tend to be brown or dark grey while males tend to be green.

venom in but if removed quickly it is unlikely that it has had time to do so. Very few people are bitten by these snakes and then mostly those that catch and handle such reptiles.

A close relative of the Boomslang, the Vine snake is a very peculiar and remarkable reptile. It is a very slender and strictly arboreal snake although sometimes descending to the ground to cross to another bush or tree. Although very large individuals may attain a length of 140 cm most snakes are between 90 and 110 cm. The females are always larger than the males, having larger heads and longer bodies but shorter tails. This can be attributed to the function of carrying the eggs, which require a large space together with the other internal organs in the slender body. These snakes are very cryptically coloured, mostly grey in ground colour with irregular black and white markings scattered over the body, while the long pointed head is blue-green above, stippled with brown. The head is distinct from the very slender neck and the tongue is red with black tips to the arms of the fork. The pupils of the eyes are horizontal and dumbbell shaped. Anterior to the eye, extending along the snout is a groove that enables the snake to see directly in front of it. In fact, it appears to have bifocal vision, as opposed to most other snakes, which can see laterally with only limited bifocal capacity. The Vine snake can therefore see stationary prey even at a distance and plans its approach accordingly. It stalks its prey approaching stealthily and watching it. If on the ground the head and forepart of the body are raised, constantly swaying from side to side with the tongue flickering as the reptile moves slowly forwards. This slow approach may be interspersed with sudden short bursts of speed. This allows the snake to approach without causing

LEFT: The head of a male Boomslang showing the large eyes, typical of a diurnal animal.

BELOW: One of the most unusual of all snakes is the Vine snake, which is highly cryptic in colour and in behaviour.

undue alarm until within striking distance. The snake then lunges rapidly forward and grasps the prey in its jaws. The poison is worked in by the characteristic chewing action and the prey is held until dead, whereupon it is manipulated by the jaws in such a fashion that it can be swallowed head first. The same method of approach is applied when the snake is in a tree attempting to catch prey underneath. When swallowing, the Vine snake may be in any position, either with the head hanging vertically downwards, horizontal or upright, none of these appearing to inconvenience it.

The Vine snake spends long periods lying in the same position in trees and shrubs, always on the alert for passing prey. Sometimes a third of the body protrudes horizontally from a bush and is held motionless, except when swaying with the wind. Its camouflage is so perfect that an inexperienced person may stand within a metre without seeing it, even when it is pointed out. In fact, it depends to such an extent on its camouflage that it remains motionless even when one encircles its body with thumb and forefinger, as long as one does not touch it. Similar head and body shape, camouflage and arboreal habits are shared with several species of colubrid snakes throughout the tropics. There are similar snakes in Madagascar, Asia, Central and South America. Some even have an almost identically shaped head. This is a classical example of parallel evolution. Three other species occur in tropical Africa.

Vine snakes are quite common in some areas. In the Nylsvley Nature Reserve in Limpopo Province a total of 147 individuals were recorded over a period of 24 months in an area of 64 ha. They occupy home ranges on average about 4,6 ha in extent, but do not display any form of territoriality. They do not appear to hibernate during the winter months, moving instead to higher ground and spending this time in dense trees. An individual may spend up to three months in the same tree or clump of trees at this time displaying typical shuttle behaviour by moving into the sun on the eastern side in the early morning, retreating to more dense and shady parts during the noon hours and moving to the western side during the late afternoon to catch the warming rays of the sun. During September they congregate in a small

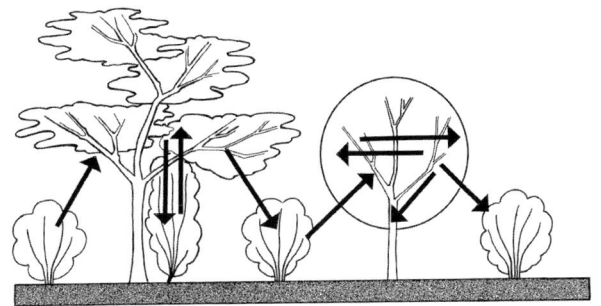

A schematic diagram of the seasonal activity of Vine snakes in the Nylsvley Nature Reserve in Limpopo Province, occurring mostly during the summer months in Sandpaper raisin shrubs, Grewia flavescens. *During autumn and winter they live in the branches of the Spine-leaved monkey-orange,* Strychnos pungens, *where they move through the tree from east to west following the sun's rays. In spring the snakes move into the raisin bushes and large Silver cluster-leaf trees,* Terminalia sericea, *where they congregate and mating takes place.*

area often inhabiting low shrubs, with up to three having been spotted in a single bush. At this time many were seen to mate, lying entwined in the branches of trees. Combat between males at this time has been recorded. They are oviparous, laying up to 18 eggs.

Their relatively inactive lifestyle seems to place few energy demands and they appear to feed infrequently, eating mainly frogs, but also lizards, birds and birds' eggs. Such eggs are swallowed whole and become softer and pliable after one or two days, due to stomach acids which attack the calcium in the shell, slowly dissolving the protective covering until the contents can be absorbed. They also feed frequently on other snakes and are cannibalistic.

They are not aggressive unless molested in which case they inflate the throat in the same manner as that of the Boomslang so that the black and white colourations are prominently displayed. They rear up and move away from the aggressor but if further molested will lunge with an open mouth and attempt to bite. They are relatively slow and have to get a good grip to chew the poison in. Bites from these snakes can be lethal as the poison inhibits the clotting ability of the blood so that excessive internal bleeding takes place resulting in death by loss of blood.

Apart from the Mole snake and Common slug-eater the only other live-bearing species in this family is the Reed snake, *Amplorhinus multimaculatus*, which produces up to

TOP LEFT: The typical colouration of the Reed snake, an uncommon live-bearing wetland snake, which occurs from sea level in the southern Cape to the Drakensberg escarpment of Mpumalanga. Photo: Richard Boycott
TOP RIGHT: A colour variant of the Reed snake, not often seen. Photo: Richard Boycott
BOTTOM LEFT AND RIGHT: The Spotted skaapsteker, common throughout most of South Africa, even in scattered localities in Namaqualand.

12 young at a time. This is an unusual and almost endemic species occurring in three disjunct populations from the Cape northwards along the escarpment mountain chain to Mpumalanga, with another in the eastern highlands of Zimbabwe. It tends to be found close to streams where it forages for frogs and lizards among reeds and riverine vegetation. It is an uncommon snake and little is therefore known of its habits and habitat preferences, and its affinities to other snakes are as yet unclear.

An interesting oviparous species is the Spotted skaapsteker, *Psammophylax rhombeatus*, which lies loosely coiled around her eggs after laying, but does not appear to incubate them. On occasion more than one clutch of eggs and brooding female may be found under the same rock. The name skaapsteker is a misnomer applied to two harmless and attractive species, which are widespread in South Africa. The name is inapt as these snakes do not bite sheep and are generally inoffensive and only rarely bite under provocation. The name was originally given to them by the early settlers who observed some mortality among their sheep due to snakebite. As these snakes were the most common they were blamed, though in reality the culprits were probably Cape cobras and Puff adders. The two species, the Striped and Spotted skaapstekers, are small to medium-sized snakes, the latter near endemic to South Africa, with relict populations in Namibia, while the former may be found from the Free State to as far north as Uganda. Their main diet consists of frogs but they will also eat lizards and small rodents. The Striped skaapsteker, *P. tritaeniatus*, has been recorded gathering in small numbers at suitable refuges during the autumn and winter, an unusual habit for South African snakes.

Among the fastest snakes are the sand and grass snakes of the genus *Psammophis*. There are eight species of these snakes ranging in size from 45 to 200 cm in length, occurring virtually in every type of habitat from desert to montane grassland, excluding only forest and ranging in altitude from sea level to 2 400 m. These swift snakes have their parallels in most other countries, indicating that under similar conditions similar types of reptiles can be expected to occur. The sand and grass snakes are our fastest species. In addition, to escape the attention of predators, their colouring is generally a nondescript brown or else they have longitudinal stripes of varying shades of brown extending down the slender body. They have a very long tail in proportion to the head and body. Some species, such as the Yellow-bellied sand snake, *P. subtaeniatus*, Jalla's sand snake, *P. jallae*, and the Cross-marked grass snake, *P. crucifer*, have yellow to orange-coloured bellies, while in most others the belly is white or cream.

When these snakes are disturbed they shoot off rapidly, but stop just as suddenly so that there is a tendency to overlook them and continue searching for them in the direction in which they fled. The stripes make it easy to overlook them when they come to a sudden stop. Other species may shoot off and just as suddenly disappear, lying quite motionless under a stone or grass tussock. Some species, such as the Yellow-bellied sand snake, may take refuge in shrubs into which they climb to avoid detection.

FROM TOP TO BOTTOM:
The Yellow-bellied sand snake often takes refuge in shrubs if pursued by a predator. It is one of the fastest of South African snakes.
An adult Cross-marked grass snake, a highveld species extending to sea level in the Western Cape.
The eggs and a hatchling of the Cross-marked grass snake.
The Dwarf sand snake, Psammophis angolensis, *an uncommon and secretive Central African species occurring in South Africa only in parts of Limpopo and Mpumalanga.*

These snakes are largely predated on by snake eagles, which generally sit on elevated perches scanning the surrounding countryside for any signs of movement.

Although snakes generally do not have fracture planes in the caudal vertebrae, some species of sand and grass snakes twist off their tail should this be in the grip of a predator, by rapidly spinning the body. In this way they make their escape leaving the predator with a section of tail. A study of caudal autotomy in this group showed that in South African species it ranged from 9,5 to 50 per cent of individuals examined. Unlike that in most lizards, the tails do not regenerate.

Their speed is not only aimed at predator avoidance but is mostly used in prey capture. All of the species appear to hunt actively, by sight. They feed mostly on lizards and some also on mice. Many of the lizards inhabiting the same areas are also swift runners making it essential for the snake to be able to put on a sudden burst of speed, and once prey is located they capture it with a swift dash.

All species of sand and grass snakes are oviparous mostly laying relatively small clutches of eggs. It appears that some species, such as the Short-snouted grass snake, *P. brevirostris*, may only have a lifespan of little more than a year, although some individuals may reach 19 months.

The Short-snouted grass snake has a wide distribution in northern and eastern South Africa. It is a short-lived species, most individuals not exceeding 19 months of age.

The urban environment has resulted in the local extinction of most snake species. Only a few colubrids have been able to adapt to such environments, including the Brown house snake, the Herald or Red-lipped snake and the Common egg-eater. These snakes occur mostly in suburban areas, but may even enter urban situations, taking refuge in rockeries, rubble heaps and other shelter especially where houses are built on hillsides. They find adequate food in mice, lizards and toads, which also frequent such areas.

THE COBRA AND MAMBA FAMILY
ELAPIDAE

We now turn from the colubrids to the family Elapidae, which is intermediate between the Colubridae and the more advanced Viperidae or adders. Its members are characterised by fangs that are short, rigid, immobile and covered by a membrane of tissue. The venom canal along the front of the poison fangs of members of this family is not completely closed, although very nearly so in some species. Some species deliver a series of rapid bites, while others bite and chew their venom in. It is disadvantageous to chew the venom in unless the prey is small and the venom quick acting. Large prey, if held for some time until the venom is forcefully injected, could seriously injure the snake. Even a mouse can inflict quite severe injuries if not rapidly immobilised. This may account for the high potency of the venom of most members of this family.

The main active principle of the venom is a neurotoxin, that is a poison which acts on the synaptic gap between the nerves to prevent the transmission of impulses across it, as well as on the central nervous system. Some species may also have a haemotoxic component, but this is of secondary importance. Symptoms include a gradual paralysis, including that of the respiratory centre of the brain causing an inability to breathe. The resulting anoxia brings about cardiac arrest.

There are numerous speculations on the potency of

different venoms and which snake is the most poisonous or lethal. There are so many different factors and facets involved that any attempt at an answer is at best a good guess. In Africa the quickest acting venom is that of the Black mamba and a bite can on average cause the death of a man, if untreated, within five hours. The locality of the bite is an important factor, as a bite received on the head and body is more severe than one on the extremities.

Virtually every continent has a snake with particularly virulent venom. Australia, for instance, has the Taipan and the Fierce snake, the former similar to the mamba, both in disposition and potency of venom. Little is known of the latter as in the past it has been confused with the Taipan, but it is regarded as being the world's most poisonous snake. Asia has the largest poisonous snake in the world, the King cobra, besides a number of others, such as the Kraits, which count among the most poisonous in the world. South and Central America also have their share, such as the Fer de Lance, Bushmaster, Jararacussu and others, while North America has Coral snakes and the Diamondback rattlesnakes. Not all of these belong to the Elapidae. A decision on which of all these is the most dangerous appears to be futile, although from all accounts the Fierce snake may well be the top contender. In spite of this, some of the most virulent poisons known may be found in a few small fish and some cone shells.

The Elapidae are widespread occurring on all continents between about 35° N and 42° S of the equator. Strangely enough both Madagascar and New Zealand are without any members of this family and apart from some back-fanged snakes there are no poisonous snakes in either of these countries, at least none dangerous to humans. North America has few species of elapids. Australia is their real home with some 65 species and the poisonous species outnumber the non-venomous ones. However, not all are dangerous. There are no vipers or pit vipers in that country, which are widespread over the rest of the world.

The elapids are a very variable group of snakes especially in size, ranging from such small species such as the Australian Dwarf crowned and Western black-striped snakes of only 25 cm in length to the giant King cobra of south-east Asia, which may reach nearly 6 m. Their habitats are just as varied, from arboreal to aquatic and fossorial, although most species are terrestrial. The Australian species are especially varied, some resembling adders in appearance and habits and are in fact called death adders on account of their virulent poison and shape. These elapids are almost all viviparous or ovoviviparous, as opposed to the African and Asiatic species which are mainly oviparous.

The South African elapids include seven genera and 17 species ranging from the small harlequin snakes to mambas. One of the largest and probably the heaviest of African elapids is the Snouted cobra, *Naja annulifera*, formerly considered to be a subspecies of the Egyptian cobra, *N. haje*, which is now recognised to be a species in its own right. The Egyptian cobra does not occur in southern Africa but is found from East Africa northwards to the Mediterranean and the Arabian peninsula, but mostly circumventing the Sahara. The Snouted cobra is a tropical species restricted in South Africa to the four northern provinces and KwaZulu-Natal.

It is probably most common in the bushveld north of Pretoria, where it is often referred to as a 'Geelslang' or Yellow snake, a name actually referring to another species, the Cape cobra. It is usually dark coloured, being slaty black above and brownish yellow below with small darker bars and streaks on the ventral scales. The head is broad and there is no noticeable neck. Some individuals are banded black and yellow, while hatchlings are yellow with a broad black band under the throat. It is possibly on this account that the snake has been given the name 'Geelslang' as juveniles of the Cape cobra are similar in colouration.

The Snouted cobra grows to a length of about 3 m and may be as much as 10 cm across the back. Despite its size it can move fairly fast over the ground and is quick to defend itself when molested, rearing up with almost a third of the body off the ground and displaying a broad hood. From this position it can strike at any foe. If anyone moves around it, it will also turn on the spot always keeping its eyes on the offender. When really aggravated it will make short rushes towards the attacker, but if not very annoyed these snakes may strike with a closed mouth, not intending to bite but merely to frighten off the source of aggravation. On one occasion a juvenile became so annoyed at being repeatedly disturbed that it actually tried to climb up a stick held in the hand of its molester.

Although often seen moving about by day it is in reality more a crepuscular and nocturnal hunter. Its prey is normally toads, rodents and frequently other snakes, especially Puff adders. Cannibalism is not unusual. At times it climbs trees and may possibly raid birds nests. Unfortunately some discover poultry and eggs and will enter chicken pens to feed on the chicks and eggs, sometimes killing adults in the process, which brings it into conflict with humans.

The Snouted cobra lives mainly in holes, especially inside larger termitaria that have established air vents to the surface, and may often be seen sunning itself outside close to the entrance. This has given rise to the Shona name 'Mungutshuru' or 'Lord of the Antheap'.

Another cobra found in South Africa is the Forest cobra, *N. melanoleuca*, which only occurs in the coastal forests of Zululand, and is a more slender snake than the previous species although almost reaching a similar length. Like the previous species it is active by day and by night. It tends to be a shiny dark yellow-brown to olive heavily speckled with black or may be a very dark olive black or even black, particularly along the posterior part of the body and tail. In contrast, the Cape cobra, *N. nivea*, is found over the Northern, Western and Eastern Cape provinces, the Free State and parts of North West Province, mostly in arid areas, and does not enter forests. Extremely variable in colour, it may range from a golden yellow to dark brown, while some individuals are brown with yellow spots or flecks on each scale giving it a speckled appearance. It is a very handsome snake with shiny scales and is an excellent rodent catcher. It is an

The largest poisonous snake in South Africa is the Snouted cobra, reaching a length of 3 m with a matching mass. Adults tend to be a dull grey-black, although banded black and yellow forms do occur. Juveniles are yellow with a broad black band under the throat.

TOP: A Cape cobra, one of our most striking snakes.
BOTTOM: The Mozambique spitting cobra will spit venom at an attacker from any position, even from ground level by tilting the head. An expanded hood in all cobras is a defensive measure carrying the message to leave it alone.

Hemachatus haemachatus. The latter is common from sea level in the Western Cape to the Free State, KwaZulu-Natal and the highveld of Gauteng, North West Province and Mpumalanga. An isolated population occurs in the eastern highlands of Zimbabwe.

Apart from injecting poison into their prey with a bite, the fangs of the spitting cobras have the exit hole situated at the front of the fang so that the venom can be forcibly ejected and sprayed forwards at its aggressor. The spitting snake usually aims high and the Black-necked and Mozambique species can reach a distance of 2,5 m, whereas the Rinkhals ejects a finer venom spray which is accurate to 2 m. The venom is harmless if it falls on the skin, but can enter wounds and cause typical symptoms of poisoning. However, if it enters the eyes it causes excruciating pain and temporary blindness within 30 seconds. Unless washed out, permanent blindness can result. The snakes do not need to rear up to spit but can do so while lying prostrate on the ground. Only the head is tilted upwards. A snake may spit as many as 10 or even more times, but with each attempt the supply of venom is further depleted, lessening the effective distance of subsequent attempts.

It is interesting to speculate on why and how this spitting habit developed, as it would seem to be an effective defence mechanism against large enemies, smaller ones being more difficult to hit. It is unlikely to have been used to procure prey. A possible explanation is that the method evolved as a protective measure against large ungulates that travel in herds. The snake would be able to spit at them from a safer distance. But then why is this habit not developed in the cobras of India, which are also exposed to the hooves of ungulates? The problem becomes still more complicated considering that the Indian cobra, *N. naja*, in the eastern parts of its range, that is in south-east Asia, is also a spitting cobra, the modification being highly developed in the region of the Malay archipelago, but is not utilised. It is therefore unlikely that the above explanation is satisfactory.

The Mozambique spitting cobra is most frequently encountered among rocky outcrops and on stony hillsides where it lies basking among the rocks near its retreat. It feeds largely on rodents but will also take toads

extremely poisonous species and may reach a length of 2,5 m. Unfortunately it too shows an inclination to raid poultry yards. Accidental bites to livestock such as sheep and other livestock also occur. It tends to be far less aggressive than the Snouted cobra and will usually attempt to flee when confronted.

There are three species of spitting cobras in South Africa. These are the Mozambique spitting cobra, *N. mossambica*, which occurs from KwaZulu-Natal northwards into the bushveld and lowveld of the northern provinces, a melanistic form of the Black-necked spitting cobra, *N. nigricollis woodi*, mostly restricted to Namaqualand, and the Rinkhals,

Another impressive and spectacular snake is the Black spitting cobra of Namaqualand. Photos: J. Marais

and frogs, besides its unfortunate taste for chickens, often being killed while raiding the chicken coop. In colour it tends to be brown to grey-brown above and pinkish below with scattered irregular black bars especially forming one large and several smaller bands under the throat. Some individuals may be almost completely black along the ventral scales. It grows to a maximum of about 1, 5 m and, like most cobras, is alert and relatively nervous so that if surprised it is quick to react and spread a hood and spit. It is, however, reluctant to bite and after exhausting its supply of venom will attempt to escape by making for its retreat. Despite this it is responsible for many envenomations in KwaZulu-Natal as it enters huts at night through openings in doors in pursuit of prey such as toads and rodents, which often find refuge and food here. In the dark people sleeping on the ground in the hut may roll onto the snake prompting it to bite. Bites by these snakes tend to cause extensive tissue damage around the site of the bite, especially if not treated within an hour. It is an egg-laying species, as are all the cobras with the exception of the Rinkhals, which is live bearing.

The remarkable melanistic colour variation of the Black-necked spitting cobra from Namaqualand differs from the normal colour of this snake, which is olive to black above and yellow to red below, but this form does not occur in South Africa, being a tropical species. The Namaqualand form, or Black spitting cobra, is distributed through Namaqualand north to Walvis Bay.

It is a large species reaching 1,8 m in length, and is black above and dark grey streaked with black below. Further north, another subspecies, the Zebra snake, *N. n. nigricincta*, a black-banded form extends into Angola.

Sweeney, in his book *Snakes of Nyasaland*, describes an interesting observation with regard to the spitting habits of the Black-necked cobra. He attempted to test the hypothesis that these snakes always aim for the eyes. Using reflecting pieces of glass held at different heights and angles, he established that in most cases the venom was aimed at the reflection. The same happened when a piece of glass was held in front of the eyes. He came to the conclusion that it was the reflection of light from the eyes that caused the snake to direct its aim. It appears to be an innate mechanism and not acquired by learning as snakes when startled may spit spontaneously and with accuracy.

The Rinkhals is a familiar and typically South African snake. However, many people confuse this species with the Mozambique spitting cobra. Both species may occur in certain areas such as in the vicinity of Johannesburg, where their ranges overlap. The Rinkhals is a variable coloured species, with the typical form black above and below, with white bars and bands under the throat, occurring primarily on the highveld and in the Free State. A banded form of narrow black alternating with white to yellow bands above and yellowish to bluish-white below occurs primarily in the Eastern Cape and KwaZulu-Natal. It is mostly a relatively short stocky snake rarely exceeding 1,2 m in length, although individuals up to 1,5 m have been recorded.

The Rinkhals is a typical grassland snake preferring areas near water, such as vleis where they feed on rodents and toads. They range from sea level to altitudes of 2 770 m in the highlands of Lesotho. Behaviourally they differ considerably from the Mozambique spitting cobra, with which it is so often confused. It is generally much less likely to spit than the latter, but some variation does occur. Its hood also tends to be much narrower when spread in a defensive position. If it finds that spitting does not deter its adversary, it will roll over gradually

The familiar Rinkhals which, like all cobras, spreads a hood if molested. However, if this defensive display is unsuccessful it will lie down and play dead.

The Black mamba is a highly nervous and timid snake which, like the cobras, may display a narrow hood and gape showing off the dark interior of the mouth if molested.

gaping its mouth and play dead. It will do this repeatedly even when righted. However, should an attempt be made to pick it up, it may spit or bite. This species is live bearing and on average 20 to 30 young may be born, although up to 60 have been recorded. A 1,58 m female from Ficksburg gave birth to 52 young. Juvenile mortality is probably high as many birds of prey, herons, storks and kingfishers feed on them, besides the many small carnivores such as mongooses and genets, as well as other snakes and even frogs.

The greatest mortality may be due to starvation as these small snakes with a total length of about 250 mm must be able to find suitably sized prey so that they can survive the severe winters experienced over much of their range. This may account for the large litters.

There are four species of mamba in Africa, to which continent they are restricted, two of which occur in South Africa. Three of the mambas are to a greater or lesser extent arboreal, while the fourth is both terrestrial and arboreal. They are found throughout Africa south of the Sahara, inhabiting tropical rainforest, savanna and scrub. They do not occur in true desert. In altitude they range from sea level to as much as 2 200 m on tropical mountains. They have a unique jaw formation with an upward curved maxilla, the anterior part of the snout being moveably attached to the brain case. The poison fangs are mostly small, about 6 or 7 mm long and situated far forward on the mandible, followed by a gap which separates them from the rest of the backward curved teeth. In addition they have elongated recurved teeth at the corners of the lower jaw, which assist them in holding on to prey while the poison rapidly takes effect. They have distinct elongated heads and slender necks, followed by a long, relatively slender body and long tapering tail. It is thought that the mambas evolved from cobras, which took to an arboreal life and became slender and elongated for swift passage through the trees. Mambas do flatten the neck and produce a narrow hood, supporting such a hypothesis.

The two species found in South Africa are the Black mamba, *Dendroaspis polylepis*, and the Green mamba, *D. angusticeps*. The former is much more widespread than the latter in South Africa occurring from the Umzimvubu River in Pondoland north to KwaZulu-Natal and three of the four northern provinces. It distribution reaches as far north as Ethiopia, Eritrea and Somaliland. The Green mamba occurs in a narrow coastal strip from northern Pondoland to Maputaland and then north to eastern Kenya.

The Black mamba is the most renowned among African snakes as it is the largest poisonous snake on the continent, reaching under exceptional circumstances a length in excess of 4 m, but is on average 2,5 to 3 m long. The name Black mamba is inopportune as its is not black but dark olive brown to brownish grey, having a greyish bloom in individuals after shedding their skin. Juveniles are generally greyish green or olive green. The underside is usually paler, with streaks of greyish brown or dark grey. The only blackish colouring found on this snake is inside the mouth, and thus Black-mouthed mamba would have been a more appropriate name.

On account of its large size, extreme nervousness and alertness, together with the potency of its venom, it is the most feared snake on the African continent. Unfortunately it has attained an undeserved reputation for aggressiveness and speed but in reality it is a very timid snake and will flee from disturbance. Even when caught by the tail it will attempt to escape, twining around bushes in an attempt to loosen the grip on its tail. If unduly molested and cornered it will rear up with the head held intermediate between that of cobras and that of adders, and open its mouth. If highly irritated it will spread the narrow hood and strike at any sudden movement, even that of leaves which have suddenly been stirred up by a breeze. The strike is slower than that of the Puff adder. It is reputed to be extremely aggressive when in flight and prevented from reaching its retreat. In an account by the herpetologist Richard Newbery, who pursued a Black mamba in the former Transvaal and inadvertently got himself into such a position the snake slid between his legs and disappeared. If pursued and unable to find a hole in the ground it will climb a tree and may be difficult to dislodge. In one instance the snake went round and round the tree canopy despite being chased, until it took refuge at the end of a branch, which was cut off and the chase continued on terra firma. Its speed of movement has been grossly exaggerated and, although fast moving, it is not as fast as sand snakes, probably due to its length. It may attain a speed of 8 to 10 km/h. Stories of a mamba biting horses in succession as it moved past them are simply that, just stories.

The Black mamba in South Africa is mainly terrestrial but will frequently climb into trees near its hole and lie basking among the branches. If disturbed, it will quickly slide down the tree and enter its hole. This may be a disused springhare or antbear burrow, a crevice between and under rocks or in a hollow tree. These snakes commonly inhabit termitaria, particularly those of the large *Macrotermes* termites, using the ventilation system as a refuge. Vacant Hamerkop, *Scopus umbretta*, nests may also be used and on occasion even the roofs of houses may be occupied.

It is a diurnal snake, and although it may sometimes move about on a warm summer's night, it is usually found outside its hole basking or foraging between 09h30 and 16h00. Food consists mainly of rodents, squirrels, birds and even bats. It is frequently mobbed by squirrels whose alarm calls reveal its presence. One of the Shona names for this snake is 'muRovambira' meaning dassie killer, on which it is recorded to feed on in South Africa and Zimbabwe, no doubt taking young individuals. This snake is indeed frequently found in rock jumbles, which are also the home of dassies. Unlike cobras, which to some extent may chew in their venom, the Black mamba bites and releases its prey to await the death of its victim before locating it and nosing it prior to swallowing, which is usually head first. The venom is so potent that the victim seldom goes far from the site where it was bitten.

The Black mamba is oviparous, laying up to 14 eggs which hatch after an incubation of about two and a half to three months. The hatchlings are 37 to 50 cm long and if sufficient food is available, grow rapidly. A hatchling reared in captivity grew to a length of 1,6 m in one year although this rate may not be achieved in the wild as feeding may be seasonal and food not so readily available. Another hatchling reared by Hennie Erasmus of Pretoria grew to almost 1,8 m in a year. It is probable that initial growth is rapid, but slows down with age. Black mambas have lived in Zoological Gardens for almost 10 years and when adulthood is reached it is not unreasonable to assume that this could happen in the wild as well, as few predators would attempt to attack snakes of 3 m in length. They have few enemies, although other snakes as well as snake eagles, Secretary birds, mongooses and genets may feed on the young. Very little is known about diseases in the wild, but there are external parasites such as ticks and mites, and internally nematodes and protozoan blood parasites have been found.

The Black mamba is no longer as common as it used to be, for it is a species that does not adapt to disturbances and is always in conflict with human interests. It may still be relatively plentiful in suitable habitat in remote areas, but such areas are becoming scarce. It is important to leave this unique snake alone in its natural habitat so that people may always be thrilled at the sight of this 'king' of African snakes instead of seeing it in a glass cage where its natural gracefulness and beauty is lost.

The Green mamba is one of the jewels among snakes with its iridescent green colour above and yellowish green belly. It has the typical mamba-shaped head but the inside of the mouth is white to bluish white. This snake is usually seen as a bright green streak gliding through the trees illuminated by patches of sunlight. It is very agile in the trees, crossing open spaces between branches with ease. This snake rarely attains a length of more than 2,2 m and is usually between 1,7 to 2 m long. It is a very shy retiring snake that does not open its mouth on confrontation. Although its venom is highly toxic it is less so than that of the Black mamba. It feeds mostly on birds and chameleons, but may include arboreal rodents in its diet. Like the Black mamba, the Green mamba also has elongated teeth at the front of the lower jaw, which probably assists in holding on to small prey such as birds and chameleons, as it might be difficult to follow up a bitten bird if it still managed to flutter away. The species is likewise oviparous, laying up to 10 eggs. The juveniles are a bright bluish green with a slightly rounded head, which later changes to the typical elongated shape. These snakes are often confused with the Green snakes and Boomslang, particularly in the northern provinces, but as mentioned before they are restricted in distribution to a narrow coastal strip.

Among the remaining South African elapids are some truly unique and beautifully coloured snakes. The Shield-nosed snake, *Aspidelaps scutatus*, and Coral snake, *A. lubricus*, are two very interesting species, about which relatively little is known. The former has the rostral scale at the tip of the snout greatly enlarged. This appears to be an adaptation to digging, which it does by pushing through the soil and leaf litter and at the same time hooking sideways at the soil with the overlapping edges of this scale. Similar adaptations occur in snakes from Asia and North America. The snakes are closely related to the cobras and rear up, but do not spread a hood, instead they inflate the throat and body and hiss loudly by expelling this air. Perhaps because they are relatively sluggish they are easily aroused and will strike out at the source of disturbance, often with a closed mouth, attempting to frighten off the intruder.

It is a small snake with a thick, stout body, rarely exceeding a length of 60 cm. It feeds mainly on rodents and toads and possibly other reptiles as well. It is normally found in sandy areas and has a wide range in southern Africa from northern Namibia eastwards across Botswana to south-western Zimbabwe, southern

Another relative of the cobras is the Shield-nosed snake, which is quick to take offence when molested, raising its head and with much huffing and puffing, strikes out at the source of annoyance.

Amongst our most beautiful snakes are the harlequin snakes, of which the Spotted harlequin snake with all its colour variations ranks the highest.

Mozambique and Mpumalanga, Limpopo and North West provinces in South Africa.

The Coral snake is a western and southern species, occurring from the Cunene River in the north to the Albany district of the Eastern Cape Province. It is more slender than the previous species, and is conspicuously banded with alternating bands of black and orange to coral red. There are from 25 to 49 bands of which the first forms a collar just behind the head. This snake favours sandy and rocky terrain in open arid country. Being nocturnal, it only emerges after dark to seek its prey, which consists mostly of snakes and lizards, although it also feeds on small rodents. Like the Shield-nosed snake it will rear up and hiss loudly if confronted and although venomous bites are rare, and then mostly to people handling them.

The remaining elapids include the harlequin snakes, garter snakes and sea snake. There are two species of harlequin snake, the Spotted harlequin snake, *Homoroselaps lacteus*, and the Striped harlequin snake, *H. dorsalis*, are endemic small front-fanged snakes rarely exceeding a length of 60 cm although the latter only reaches 30 cm in length. Both species are beautifully marked, with several regional colour variations. The Spotted harlequin snake is especially attractive and may be yellowish white above with more or less broken transverse black bands throughout. Often a bright orange to red vertebral stripe extends down the back from the head to the tail tip. The underside may be yellowish white with a dark median stripe or it may be totally black (southern and western variation). Another colour variation is black above with irregular yellowish crossbars and a series of orange to red spots or a continuous stripe down the middle of the back and reticulate bright red and yellow markings on the side of

The Striped harlequin snake is a rare species with scattered disjunct populations, some of which are threatened with extinction.

the body. The head is black and spotted with yellow, the underside barred with black and yellow (south-eastern variation). The throat is pale. A third form is black above, but with each scale having a yellow spot and with a bright orange yellow streak over the entire length of the back and tail. The head is black with yellow spots, the underside yellow with narrow dark brown to black crossbars (northern variation).

Although front fanged, these snakes are small and unable to open the mouth wide enough so that bites are rare. They tend to be found under stones or in decaying termite mounds, and are uncommon although widespread. They feed primarily on burrowing lizards and blind snakes as well as termite larvae and eggs. They are oviparous, laying six to nine eggs although larger clutches are known.

The Striped harlequin snake is much more uncommon than the previous species and is listed as Rare in the South African *Red Data Book*. It exists in four disjunct populations in the southern Free State, the highveld of Gauteng, Limpopo and Mpumalanga provinces and the KwaZulu-Natal midlands, occurring in grassland at altitudes from 1 500 to 1 800 m where it inhabits decaying termitaria. It is likely that with the extensive urban and suburban development in Gauteng, that this population is being threatened with extinction.

The garter snakes are considerably larger than the harlequin snakes and again only two species occur in South Africa, but have wider distributions beyond our borders. These include Boulenger's garter snake, *Elapsoidea boulengeri*, and Sundevall's garter snake, *E. sundevallii*, both species being similar in colour. The latter is comprised of several subspecies mostly limited to the eastern parts of the country. Variations in colour include dark stripes alternating with silvery white to brownish orange ones, plain blackish above and brownish with brownish orange below, or they could be grey above and whitish below. They have depressed shovel-shaped heads and the tail is relatively short. They are mostly found under rocks or in moribund termite mounds but appear to be nowhere common, emerging mostly at night to forage and feed on lizards, snakes (they are partly cannibalistic), frogs and also on rodents and moles.

Any discussion of the family Elapidae would be incomplete without a mention of that peculiar and unique group of snakes that have returned to the sea. Formerly considered to be in a family of the their own it is now considered that the sea snakes only form a subfamily of the Elapidae. Their fang location and structure and the nature of the venom are similar to that of other members of this family. They are wholly or almost totally dependent on the sea and are therefore highly aquatic species with a number of specialised modifications for this specific lifestyle. The body is strongly depressed dorso-ventrally or laterally, forming an oar, which facilitates propulsion through the water. Being air breathing they must surface periodically. The

Sundevall's garter snake is typical of highveld grasslands but is uncommon throughout its distribution.

nostrils are therefore situated on top of the snout and fitted with internal valves, which can be closed when the snake is submerged. The scales are small and round but still overlap, while the ventrals are much reduced in size. All species, with one exception found in Lake Taal in the Philippines, are marine and either live on the seashore making excursions into the sea, or remain in the sea for the duration of their life. They feed on fish and have extremely toxic venom, but as most species are inoffensive few bites occur. About 50 species in 16 genera occur in the Indian and Pacific oceans, mostly the latter, but only one, the Yellow-bellied sea snake, *Pelamis platurus*, occurs in water along the east coast of South Africa. This is a very widespread species extending virtually across both oceans. However, along the shores of South Africa it is a relatively rare species, with only an odd individual being washed up on the shores. Most sea snakes prefer shallow seas, such as the waters around the Philippines and in particular the estuaries of Malayan rivers. Only the Yellow-bellied sea snake appears to be pelagic drifting with the ocean currents, and basking on the surface when the opportunity arises. It is mostly 60 to 75 cm long but may in exceptional cases reach a metre in length. Like most sea snakes the Yellow-bellied sea snake is live bearing producing up to 18 young.

THE VIPER AND ADDER FAMILY
VIPERIDAE

The family Viperidae is generally considered to have the most advanced fang structure and is divided into two subfamilies, the Viperinae or adders and vipers, and the Crotalinae or pit vipers and rattlesnakes. The former are distributed widely in Africa, Europe and Asia while the latter occur in Asia and the Americas.

Members of the family are characterised by having a pronounced head, which is distinctly broader than the neck. The body is thick and mostly stocky and the tail is relatively short to moderate in length. There is usually a marked sexual dimorphism in tail length, females having short tails and males longer. The large head scales, typical of the other families, have been replaced by small scales. The long poison fangs and replacements are attached to the short movable maxillary bone, and normally lie folded along the palatine and pterygoid bones, only being raised when striking. This family is mostly terrestrial, but several species are arboreal. These snakes have adapted to a wide range of habitats from desert to rainforest with a considerable radiation in the dwarf montane adders of South Africa. The thickset body appears to be an ancient characteristic, which brings with it certain limitations. These snakes do not actively hunt prey but wait in ambush, capturing prey as it passes. The large heavy body is ideally suited to a terrestrial mode of life and it is consequently surprising that some viperids have become arboreal, developing a prehensile tail to assist in climbing. All species can swim and one, the Water moccasin or Cottonmouth, belonging to the subfamily Crotalinae, living in swampy country such as the Everglades in the American state of Florida, is semi-aquatic. It resembles the Rinkhals, but the latter species does not occur in such swampy terrain. The nature of the poison found in adders, pit vipers and rattlesnakes varies considerably from cytotoxic to cytotoxic and neurotoxic, while others are haemotoxic. Many species have venoms dominated by one or other of these types, but incorporating components of the other types. Side effects from bites by these snakes are common, including gangrene.

Many viperids feed on warm-blooded prey, mostly rodents. Most species lie in wait striking out at passing prey. The relatively long fangs and poison glands of this family have probably developed to accommodate the type of poison found. This is not as quick acting as that of the Elapids, so that the long fangs deposit the venom deep for maximum efficacy, as well as causing considerable tissue damage at the same time. Consequently these snakes usually bite their prey and leave it. After a while the snake will follow the scent trail and locate the prey, now dead and unable to fight back and perhaps injure the snake.

There are about 150 species in the family within some 14 genera. About 100 species belong to the Crotalinae and 50 species to the Viperinae. The former are divided into five genera and the latter into nine. None of the Crotalinae occurs in Africa. As mentioned previously, the adders and vipers are restricted to the Old World with one species, the European adder, *Vipera berus*, extending as far north as the Arctic Circle. However, most species live within the tropics and temperate zones north and south of the equator. The genus *Vipera* is mostly European and Asian, with approximately seven of its 10 species found in Europe. Only one species has penetrated the south-east Asian tropics, namely Russell's viper, *Vipera russelli*, a greatly feared species. The other vipers live in Central and south-western Asia. The European adder or viper, is the only poisonous snake in England and is found in all European countries except Ireland, which has only one reptile, namely a lizard. This snake also has an extensive distribution outside Europe extending across the temperate zones of Asia.

The deserts of Africa, central and south-western Asia are the home of several genera of adders, which are very diverse but similar in habits. One such species is the Saw-scaled viper, *Echis carinatus*, which occurs from northern Kenya westwards to India and Sri Lanka. It may occur in tremendous numbers in some parts of its range such as north-western India where more than 200 000 were killed annually over a period of six years. It is an interesting analogy to the egg-eaters as it produces a hissing sound by rubbing the keeled scales on opposing sides of the body, which is usually formed into a figure of eight, against each other. It is responsible for a great many bites and coupled with a very potent poison is considered one of the most dangerous snakes. Others include the horned vipers and the false horned vipers of North Africa, of which there are two species each. These snakes have valve-like structures inside the nostril and the lips also appear to seal tightly, possible modifications against windblown sand in a desert environment.

The continent of Africa and South Africa in particular is the home of adders of the genus *Bitis*, which include a great diversity in size and adaptations. Although restricted to Namibia and southern Angola, mention must be made of Peringuey's adder, *Bitis peringueyi*, a small snake rarely reaching 45 cm in length and thick-bodied. Just as the American sidewinder, *Crotalus cerastes*, is adapted to life on shifting sand dunes, so is this adder. Its head is depressed and the eyes are situated on the top of the head so that when it is buried in the sand only the eyes protrude above the sand surface. This burrowing habit is accomplished by the lateral movement of the body raising the ribs on the outside, which acting as a scoop pushes sand away and at the same submerging the snake. A return action removes sand on the inside of the body, so that ultimately by alternately scooping inside and outside the animal is buried in the sand. This camouflage behaviour not only serves to conceal the snake while it waits for passing prey, but also protects the animal from predators and, more importantly, is a thermo-regulatory mechanism as lying exposed to the sun would kill a reptile very quickly. The snake also has a black-tipped tail, which may be exposed and held up while it is wiggled about when prey approaches. It has been suggested that this serves as a lure distracting the attention of prey so that it may come within striking range. Normal prey consists of swift-running lizards, which when pursued may dive in the sand and disappear. This makes it impractical for Peringuey's adder to chase or follow the prey after a bite so they are gripped in the mouth until the poison takes effect.

Peringuey's adder is the typical sidewinder of the sand dunes in the Namib.

TOP: *Another arid adapted snake is the Horned adder which has a wide distribution in South Africa.*

BOTTOM: *The irascible Berg adder occurs in disjunct populations along the escarpment mountains extending down to sea level in the Cape.*

This snake, like that of the Sidewinder, has a peculiar mode of progression. Peringuey's adder progresses by lateral undulations of the body during which the curves of the body are lifted and moved sideways in a series of even and continuous loops. This method of progression is essential in loose sand, leaving tracks in the form of parallel grooves lying across the direction of travel. Another species, Schneider's adder, *B. schneideri*, which is endemic to the Namaqualand coastal dunes, is similar in habits and appearance but is a rare species occurring mostly in the diamondiferous areas along the north-west coast.

Some other adders also have adopted this mode of progression but is not utilised to the same extent, as these species also inhabit more stable terrain. One of these is the Horned adder, *B. caudalis*, a common and widespread species occurring from northern Namibia to the southern Cape and through the Kalahari to the North West, Gauteng and Limpopo provinces and into western Zimbabwe. This snake is characterised by elongated horn-like scales above each eye. They inhabit a variety of habitats, among others occurring along the Magaliesberg, where they may be locally common. They are short, thick-bodied snakes rarely exceeding 30 to 37 cm in length, exceptional individuals reaching 50 cm. In areas of deep sand such as river beds, they frequently bury themselves in the sand in the same way that Peringuey's adder does. In addition they also twitch and lash the tail tip, which protrudes above the sand. They feed mostly on lizards and mice, and are viviparous with as many as 18 young being produced at a time.

The mountain adders occurring along the mountains from the southern Cape to the Limpopo Province, with a relict population in the eastern highlands of Zimbabwe, were originally lumped as a single species. It has now become apparent that what was thought to be one species is in fact a complex of five species, four of which are endemic, inhabiting the mountains of the southern and eastern Cape. The fifth species, the Berg adder, *B. atropos*, occurring further north along the KwaZulu-Natal Drakensberg, parts of Mpumalanga and Limpopo provinces to Zimbabwe. The most northerly population of this species in South Africa is at Haenertsburg. Populations appear discrete and widely separated by unsuitable habitat. Such populations are at risk due to habitat destruction by afforestation and firebreak burning practices. The population at Haenertsburg is one of the most threatened. Numerous individuals with scorch marks down the middle of the back give testimony to the effect of such fires. The Berg adder is an attractive snake, dark brown with dorso-lateral sub-triangular to semi-circular black markings on either side. It is edged with a white line below. Various shaped blotches and markings make this one of our most striking adders. They are small, averaging 30 to 40 cm in length, rarely reaching 60 cm. They also give live birth, the young varying from eight to 15 depending on the size and age of the female.

These snakes are irritable and individuals may warn an intruder of their presence from a distance of 5 m by hissing loudly. They frequently lie in paths and on grass tussocks basking in the morning sun. If disturbed they wriggle away quickly into thick grass cover but bite readily if provoked. Their venom resembles that of Puff adders but includes a neurotoxic component. Bites are usually not fatal, but cause considerable unpleasantness including temporary partial blindness, loss of taste and hearing impairment. They feed mainly on lizards and mice.

The Puff adder, *B. arietans*, is our most common venomous snake and is found from sea level to mountain tops at altitudes of 2 000 m. It ranges virtually throughout South Africa and extends north to Arabia, occurring in most habitats with the exception of true desert and rainforests. It varies greatly in colour from region to region usually in relation to the habitat and soil colour of the area in which it resides. In the southern and eastern Cape specimens are dark and spotted with yellow with yellow-orange under the throat. Elsewhere it is varying shades of brown to blackish, with chevron type markings along the back. It is this cryptic colouration coupled with its habit of lying still, even when it feels something approaching that makes it one of our most dangerous snakes. It relies on its camouflage to remain undetected and therefore out of harm's way. Puff adders are medium-sized snakes but heavy bodied. They average about 75 cm in length but may reach 1,2 m, although in East Africa they are reputed to reach up to 1,5 m, with a proportionate girth. An average Puff adder weighs anything between 0,5 and 1,2 kg. Growth can be fairly rapid and a captive male reached a length of more than a metre in seven years. It also appears that once a length of 84 cm is reached, the snakes begin to increase in weight faster than they increase in length. A male was still growing after 11 years and eight months with a length of 109,5 cm and a mass of 1,535 kg.

On account of their stout bodies they move in a straight line by using their belly scales and muscles attached to the ribs, each scale gripping the surface and propelling the animal forward, in a similar motion to that of millipedes. However, if molested they thrash about and then break into a clumsy serpentine movement. This cannot be maintained for any length of time before seeking shelter. If approached they hiss loudly, as the name suggests, with violent expulsions of air, when really agitated. This is purely a defensive and warning measure, which when ignored results in the snake adopting a threatening strike position with the body partly coiled and the head held high and arched on the neck. From this position they can strike with lightning speed. Like all of the adders the strike is very fast, much faster than that of other African snakes.

The direction of the strike is always to the front or side but not behind and it certainly cannot strike backwards. Small individuals may strike so violently at times that the body lifts off the ground for a moment,

Probably the best-known snake in South Africa is the Puff adder. The relatively long tail is typical of males.

A Puff adder in a defensive position from which it can strike forward at great speed. The short tail indicates that it is a female.

A rare scene, male Puff adders in combat, one attempting to assert dominance over the other, usually in order to mate with a female in the vicinity. Photo: Richard Boycott

but the snake does not leap forward as has been claimed.

The fangs of this snake are large in order to inject the poison deep so that it rapidly takes effect. A Puff adder of one metre in length may have fangs 20 mm long, but these may even be larger in the large East African individuals. The venom is highly potent and the victim may die depending on the treatment and other factors. It is haemolytic in action breaking down the red blood cells, as well as the walls of capillaries, while a general necrosis develops around the site of the bite. The latter is very painful and swelling is very pronounced, a fact which is used to distinguish the bite of this snake from that of cobras and mambas.

The normal prey of the Puff adder is rats and mice, particularly species such as Vlei rats, *Otomys* spp., Striped field mice, *Rhabdomys pumilio*, and Multi-mammate mice, *Praomys natalensis*. These rodents move about along runs established in thick vegetation and are easily captured by the snake lying in ambush. The snake usually does not retain a hold on its prey but releases it to avoid injury to itself. The victim may be able to run away for a short distance before the venom and the mechanical injury inflicted by the bite takes effect. After a short while the snake follows the scent trail locating its prey with great accuracy. Once found, it noses the prey from head to tail flicking its tongue along the length of the body. This serves to identify the location of the head. It then manipulates its jaws over and under the prey, the long fangs assisting by hooking into the prey and bringing it into the mouth, where the palatine and pterygoid teeth take over helping to pass the animal into the throat where the neck muscles manipulate it down the gullet. Puff adders are excellent rodent catchers and are extremely important in helping to keep population levels of these animals down.

Similar to the other adders, Puff adders give birth to

20 to 40 young and there are records of up to 60 or 70, but this must be considered exceptional. The size of the brood is partially dependent on the size of the female. Puff adders reared in captivity have first given birth at the age of three years and nine months, having attained a length of almost a metre. The young are fully developed at birth with a functional venom apparatus and are easily irritated at this stage, striking out freely. There is no truth in the tale that the young snakes eat a hole in the side of the female, which probably had its origin in a snake that had been killed and had a hole in its side from the injuries it had received. The young, fully developed, were able to wriggle free of their covering membrane and exit from the mother via the hole in her side.

Similarly there is no truth that the young take refuge in the mother's mouth. This story is widespread, actually originating in England, where a writer conferred this habit to the European viper in the year 1577. This story travelled to this country, similar habits being ascribed to similar snakes.

Puff adders can be long-lived, potentially up to 20 years, but the longest authentic record appears to be 13 years and 11 months. It is difficult to ascertain how long these snakes live in the wild but it is doubtful if they ever reach this age.

While the Puff adder may well be the best-known adder, the Gaboon adder, *Bitis gabonica*, is certainly the largest and most striking. This giant may reach a length of 1,8 m in tropical Africa, but usually reaches about 1 to 1,2 m in South Africa and weigh up to 8 kg. The head is very large and triangular with a thin neck and thick body. It is beautifully marked with pastel colours, which blend well with its surroundings.

The Gaboon adder is a very sluggish and inoffensive snake, in spite of the enormous 50 mm long poison fangs in a 1,8 m individual. It must be considerably provoked

A rare endangered species in South Africa is the Gaboon adder. The recent destruction of much of the Dukuduku forest in Zululand by squatters has substantially increased the risk of extinction of the species in South Africa.

before retaliating. There are accounts of the docility of this snake, one being of an African youth dragging a 1,4 m live snake to the camp of a Mr Herbert Lang, for the purpose of selling it to him.

They are very local and rare in South Africa, only occurring in suitable habitat such as the Dukuduku forest in northern Zululand, parts of Mozambique, eastern Zimbabwe, Malawi, north-eastern Zambia to Tanzania, west to Angola and further north. In Zululand it is found mostly along forest margins and in thickets, depending on the season, as well as utilising moist grasslands, inhabiting large home ranges with core areas of greatest activity.

Its venom is highly potent and because of the length of its fangs, which are the longest on any snake, it can inject the poison very deep and in large quantities, making it an extremely dangerous snake. Like the Puff adder, the Gaboon adder is mainly a crepuscular to nocturnal snake, feeding mostly on rodents such as Vlei rats, *Otomys* spp., Red veld rats, *Aethomys chrysophilus*, Woodland mice, *Grammomys dolichurus*, and other species, as well as birds, and rarely other vertebrates such as toads and frogs.

Other well-known adders include the night adders of which we have two species. The most common is the Common or Rhombic night adder, *Causus rhombeatus*, which is encountered in varying habitats, but shows a preference for higher rainfall areas and moister sites. It is widely distributed from the western and southern Cape northwards along the eastern half of South Africa to East Africa and across to West Africa. The other species is the Snouted night adder, *C. defillippi*, which is more restricted in distribution, mainly confined to Limpopo and Mpumalanga provinces as well as northern KwaZulu-Natal, also occurring extensively north of our borders.

Both species are varying shades of brown in ground colour mostly with dark blotches extending down the back and with a forward-pointing 'V' at the back of the head. The Snouted night adder is characterised by the upturned scale at the tip of the snout, but is generally much smaller, reaching a total length of 45 cm whereas the Common night adder may attain a length of 90 cm.

The Common night adder is a snake mostly occurring in the higher rainfall areas of South Africa.

Both species are to a greater or lesser extent crepuscular or nocturnal. Although comparatively inoffensive, both species make a great show by hissing loudly and if highly irritated will coil up the body, inflate it with air and strike out in all directions. They make such a show of this that they are sometimes referred to as Demon adders. In captivity they become quite docile and can even be handled freely. Despite being relatively common in parts, our knowledge of such species remains limited.

They feed mainly on frogs and toads, which are bitten and held firmly until dead, after which the typical head orientation of the prey is performed and it is swallowed. The poison fangs are small, about 4 to 6 mm long. The venom is not as toxic to humans as that of the other adders and is seldom fatal, but is very effective on toads. The most peculiar feature of these snakes is the size of their poison glands, which stretch far back on either side of the neck. In large individuals, they may be up to 50 mm long. However, little poison is produced, proportionately not much more than that of the other adders.

These snakes are egg-laying, with up to 25 eggs being laid by the Common night adder whereas the Snouted night adder only lays up to six. There may be two clutches a year. Females of the former species can retain viable sperm in their oviducts so that they do not need to mate more than once in a season. However, for each successive clutch of eggs the percentage of infertile eggs increases.

Crocodile basking and thermoregulating. Note the flap at the entrance to the throat, which prevents water from entering the lungs. Photo: J. Marais

CROCODILES

ORDER CROCODILIA

We now come to the crocodilians, which are probably some of most imposing of the orders of living reptiles. They bring us back to that great era of reptiles – the Mesozoic, when these creatures ruled the earth. The crocodiles, alligators, caimans, false gavials and gavials are the last surviving representatives of the age of dinosaurs. In structure and probably also in habits the crocodiles have altered little since this era ended some 70 million years ago.

The name Crocodile originates from the Latin Crocodilus, *which is very close to the Greek* Krokodeilos, *both meaning lizard. Similarly the name Alligator is derived from the Spanish* El Lagarto, *which also means 'the lizard'. The reptile order of Crocodilia consists of a single family Crocodylidae, incorporating all crocodilians. There are 12 crocodile, a dwarf crocodile, a false gavial, ghavial, three caiman, two dwarf caiman and two alligator species distributed around the tropical regions of the world. Crocodiles are widespread in distribution occurring in Africa, Asia, New Guinea, Australia and North, Central and South America, but only a single species, the Nile crocodile,* Crocodylus niloticus, *is known from South Africa.*

128 • REMARKABLE REPTILES

The Crocodilian Family
Crocodylidae

The crocodilian body is well adapted to its aquatic existence. Both the eyes and the nostrils are placed high on the head so that they can protrude from the water while the animal cruises about looking for possible food without becoming conspicuous, but still able to breathe. The nostrils are opened and closed by special muscles, which seal them when submerged. The nasal cavities are drawn into long tubes, surrounded by bone, which open far back in the throat anterior to the opening of the trachea. This region can be shut off from the rest of the mouth by valves allowing the crocodile to keep its mouth open under water without inhaling water and drowning.

Crocodilians are air-breathing animals, periodically needing to surface to obtain oxygen. They may survive lengthy periods under water, up to an hour, without drowning, but this depends on the size of the animal and it is unlikely that this occurs with any frequency under natural conditions.

The eyes are golden coloured with a vertical slit-like pupil typical of nocturnal animals, but this expands under decreasing light intensities becoming round in the dark. They are protected by a nictitating membrane when the crocodile submerges. Sight is good and Zdenek Vogel concluded that the reptiles could see the outlines of large immobile objects clearly at a distance of 5 to 10 m and movements up to 33 m away.

On land crocodilians progress in what is termed the 'high walk' in which the body is raised off the ground, with only the tail tip dragging. Mostly however, the animals slide along on their bellies,

Crocodiles are well adapted to an aquatic existence. Note the nostrils on top of the snout and the ear opening immediately behind the eye.

propelled by their feet, until they are in the water, where they use their muscular tail to propel themselves rapidly into deeper regions. Other methods used on occasion include the 'belly run' and the 'gallop', the latter mostly restricted to juveniles, and reaching speeds of 11 to 12 km/h.

The ears of crocodilians are protected by scaly flaps, just posterior and a little above the eye. These flaps can be raised or lowered, exposing the eardrums. Normally they are kept shut, with the exception of a small slit at the front, which opens when the animal's head is out of the water. They are presumably deaf when submerged.

The crocodile's jaws are formidable and armed with an array of sharp-edged, pointed, peg-like teeth, approximately 66 in number. These teeth are replaced throughout life and it has been estimated that a crocodile measuring 4 m may have had as many as 45 sets of teeth in its lifetime. The teeth are adapted for gripping and holding onto prey, facilitated by strong muscles to close the jaws, which virtually act as a locking mechanism. This is important for the capturing of large prey. The muscles that open the jaws are relatively weak and it is alleged that a man can hold a crocodile's jaws shut with his hand. It is not wise to try this as crocodiles may bite very quickly when annoyed!

As mentioned earlier, the Nile crocodile, *Crocodylus niloticus*, is the only species from this family that occurs in South Africa. It is a large animal reaching a length of about 5 to 6 m and may weigh up to 1 000 kg. This is a far cry from the giant 11-m crocodile *Deinosuchus* or Terror lizard, which lived during the Cretaceous period in North America, or *Sarchosuchus* a large prehistoric crocodile from Africa and Asia, which attained a similar size. Nevertheless it is an impressive and formidable animal.

The Nile crocodile was widespread in South Africa a hundred years ago, but the arrival of the white man, armed with modern guns and the demand for skins by the fashion industry, has almost resulted in their disappearance from their usual haunts. Their resultant decline and the establishment of many crocodile farms have largely ended this persecution, and they are a protected species in South Africa today.

Originally found along the east coast as far south as East London, crocodiles extended inland only in the northern provinces, where they inhabited the main east-flowing river systems. Although currently much reduced in numbers, they may still be found along these rivers today. These include the Olifants, Crocodile, Sabi, Letaba, Levubu and Limpopo rivers, while in KwaZulu-Natal they are still to be found in most of the main rivers in Zululand north of the Tugela River. It is doubtful that the total crocodile population in the wild in South Africa exceeds 8 000 individuals.

Dr Hugh Cott did extensive research into the habits of the Nile crocodile from 1952 to 1957 and it is due to him that we know so much about its way of life. Subsequent work by others such as Tony Pooley, formerly of the Natal Parks Board, John Loveridge of the University of Zimbabwe, David Blake and Jon Hutton, formerly of the Zimbabwe Department of Wildlife Management, has given us substantial insight into this formidable predator.

From an analysis of stomach contents of a large sample we know that the young animals feed mainly on insects and other invertebrates, but as they get older frogs and then fish form the main component of their diet. This is gradually replaced by reptiles, birds and mammals, so that by the time the crocodile is 4 m long, it feeds largely on mammals, although in many areas fish may still form the main component of the diet.

Crocodile skull and teeth. There is no cranial kinesis, the solid head is an adaptation for the capture of large prey.

Crocodile attacking. Photo: Lorna Stanton

Crocodile burrow along the Limpopo River.

Crocodiles, especially larger individuals, ingest pebbles to assist with digestion and to serve as ballast, enabling the animal to remain steady in the water or even to rest on the bottom.

There remains some controversy as to the role of the crocodile in nature and about which fish it feeds on. It appears that the Nile crocodile feeds to a large extent on fish such as barbel and lungfish, which themselves prey on the young of highly edible fish such as kurper and in this manner maintains a beneficial balance between fish species.

Crocodiles do not feed continuously and can go without food for long periods. For instance, two crocodiles in the London Zoo, each 2 m in length, had an average daily intake of only 365 and 280 g respectively and at this rate it would have taken them about 124 and 160 days respectively to consume their own weight in food. Cott has calculated that adult crocodiles do not consume more than 50 meals per year.

Crocodiles have devised various techniques to obtain prey. Young fish, for instance, may be herded into shallow water, the reptile approaching them broadside until able to rush in and catch a mouthful. When catching mammals, the crocodile gradually approaches the animal as it drinks, often at a tangent, and then rushes in and grabs it by the snout, dragging the prey under water where it is drowned. In some instances the crocodile will even rush out of the water to catch its prey. A crocodile approaching a drinking dove did this so gradually as to be almost imperceptible, only the eye ridges and nostrils protruding out of the water. After manoeuvring to within striking distance, the crocodile surged rapidly forward, propelled by its tail. On this occasion it miscalculated and the dove flew off unharmed. There are records of crocodiles using their tail to sweep prey into the water but this method is more the exception than the rule. Large prey includes rhino and hippo calves, and there are records of attacks on small elephants to the detriment of the crocodile. On one occasion an elephant cow was seen to drag a crocodile out of the water, after it had been attacked. The crocodile is catholic in its choice of food and is opportunistic.

Contrary to popular belief a crocodile is not fond of putrid meat. After taking a bite at a carcass, the crocodile rolls along its longitudinal axis, spinning and twisting meat off in this manner. To swallow the meat it has to hold its head high out of the water. It cannot chew, as the teeth are not adapted for this and chunks are swallowed whole.

Crocodiles have an interesting way of dealing with barbel. They bite the fish and with a rapid flick of the

Crocodile basking and thermoregulating. Photo: J. Marais

head jerk it about, until the heavy bony head is torn from the body and is discarded. The rest of the fish is swallowed whole.

Like all reptiles, crocodiles are ectothermic or cold-blooded animals. This phenomenon has, however, nothing to do with their blood but refers instead to their inability to regulate their own body temperature physiologically as mammals do. Like other reptiles they have to rely on ambient temperature and the sun to provided warmth, until their body temperature reaches 25 °C, at which they are capable of normal movement. Throughout the day their temperature may fluctuate around this level, varying by six degrees in either direction. This is achieved by lying on rocks and sandbanks in the morning, and retiring to lie partly submerged in the water or under bushes in the heat of the day. As they lie basking, an increase in the body temperature is compensated for by opening the mouth, the lining of which acts as a cooler, caused by the evaporation of moisture from the membranes of the buccal cavity. This enables the crocodile to maintain its body temperature even if the outside temperature almost rises to the critical maximum of 38 °C. The loss of water can amount to as much as 20 per cent of the body weight in 24 hours, but is replenished by drinking when the animal enters the water. At night the crocodile spends most of the time hunting in the water.

Crocodiles live in a wide variety of habitats including rivers, lakes, streams, dams and pans. Many of these dry up during the dry months of the year. They spend this period lying up in humid sites under overhanging banks, or under the exposed roots of trees and even in caves but mostly in holes in the banks dug by the crocodiles themselves during the dry season. Such burrows are of varying lengths, up to 12 m long but are mostly shorter, with a chamber at the end. One or more crocodiles may inhabit these chambers. Digging is with their jaws, the animals biting out chunks of soil just above water level, removing it outside the burrow where they wash the soil out of their mouths by shaking their heads. More than one animal may work simultaneously on one burrow. Burrows have been recorded from Ndumu in northern Zululand and along the Limpopo River in Limpopo Province.

As an alternative to digging burrows and hibernating, the crocodiles may leave their shrinking pool and travel overland in search of more permanent water. Similarly during the rainy season they may also travel far along tributaries or shallow drainage lines. An isolated pool at the head of a dambo in north-west Zambia some 32 km from the nearest water, was found to contain a number of small crocodiles basking around the margins of these pools. Such movements may also be initiated by young animals seeking to escape from aggressive adults, especially territorial males. Along the middle Olifants River large crocodiles mostly occurred in deep pools in the vicinity of Marble Hall while further downstream in Sekhukuniland, numbers of juveniles occupied shallow pools in the bends of the river. The juveniles obviously found refuge here, away from the cannibalistic tendencies of adult animals. Larger males tend to establish themselves in certain pools and areas and become territorial, chasing out younger, smaller males, often resulting in these travelling far up rivers to find a place to live. Though such behaviour has been documented in the wild, it becomes really prominent in captivity, particularly in the mating season, usually during July. Smaller males are then terrorised by the dominant males and retreat to the remote corners of their enclosure, where they may be tolerated. Under such conditions severe fighting can break out, and Dave Higgins, a former crocodile farmer from Zimbabwe, told of a large 5 m male seizing a 3 m individual in its jaws and shaking him much like a terrier shakes a rat. Dr U. de V. Pienaar, former Chief Director of the National Parks Board, also related a similar incident from the wild, which took place in Lanner Gorge along the Levubu River. He observed a similar sized crocodile biting and shaking an estimated 3 m crocodile.

This feat of strength may seem unbelievable, but one has only to look at a crocodile or merely handle a small one, a metre in length to appreciate their strength.

During the mating season the males may roar, with the head lifted and jaws open, producing a sound like the roll of a drum. The females respond with a growling roar. The males also bark and cough in deep hollow tones. They give off a heavy musk odour, the female more so than the male, which aids in locating a mate of reproductive readiness. Mating takes place in shallow water. The male approaches a female and by vibrating his head alongside the female, tests the response. If the female is receptive the male mounts the female twisting the hind part of his body under hers so that their cloacas come into contact. Unlike lizards and snakes, male crocodiles only have a single penis, which is inserted into the cloaca of the female.

In South Africa, the eggs are laid from November to December and the young hatch between the end of January and the end of March. This coincides with the height of the rainy season, when conditions of heat and moisture are at an optimum for development, though this can be deleterious when too much rain falls and the young drown or the eggs are attacked by fungi. The eggs are buried about 50 cm deep in a pit dug by the female. Digging is mostly done at night or in the early morning hours, although numerous observations indicate that late afternoon may also be suitable. Nests are placed in a variety of sites and can be as much as 15 m from the water.

In South Africa an average clutch of 45 eggs is laid but this may vary between 18 and 91, depending on the size of the female, but clutches over 80 are exceptional. The eggs are deposited in tiers in the pit, each layer usually liberally covered with sand, to prevent damage. The nests are dug using the hind feet alternately as a scoop to remove the loose soil and deposit it on one side.

As soon as the eggs have been laid, the female proceeds to cover them in such a way that the topmost tier is about 20 to 40 cm below the surface. The actual depth of the nest is more or less dependent on soil texture and compaction. The hole is filled with the feet and the soil is tamped down with the belly, simultaneously removing all traces of activity.

Where conditions are ideal and crocodiles are numerous, nests can be found literally all over the available habitat. Cott recorded no less than 24 in an area of 525 m² on the shores of Lake Albert in the Democratic Republic of Congo. The female leaves the nest and the eggs incubate in the warmth of the sun but she usually lies close by and protects it, often defending it from incursions by other females also seeking a place to lay their eggs, or from crocodiles looking for a place to bask. She may at times almost lie on top of the nest and even abstain from feeding. In captivity females have been recorded protecting the nest from water by moving on top of the nest. She is aggressive at this stage and resorts to chasing an intruder away by rushing with an open mouth, growling and lashing her tail. If driven off she tries to attack from another direction and when in the water, smacks the surface with her chin, producing a loud explosive sound, which can be startling in its unexpectedness.

Apart from fungi, the eggs may be subjected to drowning if covered by water for any length of time during floods, most being destroyed. Such catastrophes are by no means the only danger to the eggs, as they may be consumed by a number of animals, and regardless of how carefully they are guarded, a large percentage are eventually found and destroyed by predators during the three-month incubation. The

Adult Crocodile. Photo J. Marais

most notable is the Water monitor, a species common to our rivers, streams and dams. These large lizards, which may grow to 2 m in length, have a definite regulatory effect on crocodile populations.

Other predators include the Water mongoose, *Atilax paludinosus*, Banded mongoose, *Mungos mungo*, Spotted hyena, *Crocuta crocuta*, baboons, Warthog, *Phacochoerus aethiopicus*, and Bushpig, *Potamochoerus porcus*, while birds such as Marabou storks, *Leptoptilus crumeniferus*, and vultures also help themselves to the eggs of uncovered nests. In this manner entire clutches are decimated, with the sun, ants and scavengers completing the final destruction.

Incubation takes 11 to 14 weeks, the eggs becoming turgid and the shell softens and cracks. The hatchlings force their way through the inner integument, aided by a small egg tooth on the tip of the snout, which drops off soon afterwards. As the time of hatching approaches, the baby crocodiles utter an explosive 'eyoh!' These sounds indicate to the mother that they are hatching and she digs open the nest, scooping a round crater-like pit uncovering the top layer of eggs. This is important, as the soil by now has become compacted making it difficult for the young to dig their way out. The instinct to dig the young out is very strong and the mother will tackle almost any obstacle to perform this duty.

This was illustrated by Tony Pooley, who built a hide around a nest on the banks of the Pongola River in northern KwaZulu-Natal. It was made of 15 cm thick poles joined by eight-gauge fencing wire and closed off with hessian. At hatching time the parent smashed the bottom of the hide and tore away the sacking. A large hole revealed that the nest had been opened and the hatchlings released. Immediately after hatching the female gathers the young in her gular pouch, picking them up individually in her mouth.

Another account by Tony Pooley concerned the interesting happenings when a foster brood was given to a large female. As a live hatchling was introduced into her pen, this female lunged forward with head held sideways and attempted to pick it up with her teeth. The hatchling called several times and tried to climb through the fence but she grasped it delicately between her teeth and pulled it back. Then with a quick gulping motion manoeuvred it between her jaws and with a second upward lift of the head put it into the pouch. Over the next two hours a further 18 live hatchlings were introduced and each was captured and put into the pouch. Some of these hatchlings walked straight towards her jaws and of their own accord climbed inside as she opened her jaws. Others became excited at her approach, calling almost continuously and flicked their tails from side to side as if to draw attention. Once inside her mouth, which she held half open, not one of them attempted to escape and on closer observation several appeared to be sleeping. After an interval of five minutes from the time the last hatchling had been picked up, the female turned around and lumbered off down to the pool, situated 15 m from the fence, entered the water and disappeared into thick floating vegetation. Almost immediately the calls of the young crocodiles could be heard as they were released into the water.

Females have also been observed taking hatching eggs to the water, where the shell is gently cracked and the hatchling released. This parental behaviour is unique among reptiles and borders on that of mammals.

The hatchlings, only about 28 cm in length, are agile and lively. They remain for a time with the adult, but keep to the shallows along rivers and pools, especially well-vegetated ones. They are extremely vulnerable at this stage of life and stay close to whatever shelter is available. Away from such cover they quickly fall prey to a host of predators both on land and in the water. Monitor lizards, genets, mongooses, Marabou and Saddle-bill storks, herons, ibis, fish eagles, fishing owls, ravens, eagle owls and Ground hornbill prey on them, especially following hatching. In the water they fall prey to otters, pelicans, tiger fish, barbel, turtles and crocodiles. In fact, predation by other crocodiles may be among the most important.

Hatchlings are initially gregarious, but soon disperse feeling safer on their own and remaining silent. They do not require food at this stage as they have sufficient nourishment from a yolk sac, about the size of a hen's egg. This is gradually re-absorbed by the stomach and traces may still be visible after six months. By this time they are already catching invertebrates and frogs but it is likely that many die at this time due to an inability to find and catch suitable food. Only a small percentage, estimated at about 1 to 2 per cent of the total number of hatchlings, survive their first year of life. They disperse widely during this time, wherever suitable shallow water is found. Once they reach a length of 1,5 to 2 m they are relatively safe from predation by their own kind and tend to occur in the same pools as larger individuals.

Growth is relatively rapid during the first seven years and an average rate of 26,5 to 28 cm per year has been recorded, slowing down to 3,6 to 4 cm for the next 15 years. Growth rate is correlated to food availability. Once individuals reach 4 m, they increase progressively more in bulk than in length. Males reach sexual maturity at 2,9 to 3,3 m in length with a mass of 100 kg, and females begin breeding at 3 m in length, few breeding before reaching at least 2,44 m. The smallest breeding females recorded had a length of 2,18 to 2,33 m.

Larger crocodiles have few predators, the foremost of which is man. Although crocodiles are eaten by people to some extent and may be hunted extensively for this purpose in some countries, it is for their hide that crocodiles all over the world have almost been hunted to extinction. The Siamese crocodile, *C. siamensis*, was considered extinct in the wild, but a surviving wild population was recently discovered in remote parts of Cambodia. The Nile crocodile, despite its man-eating tendencies, has been more fortunate and is still widespread in Africa and Madagascar. There are occasional instances of elephants and hippos killing adult crocodiles in defence of their young, while leopards and lions may prey on those found on land.

Crocodiles live to a ripe old age and have been recorded living for 25 years in a zoo. In the wild observations show that after 22 years growth continues at a rate of approximately 3,5 cm per year, which would make a 4,58-m individual about 76 years old and larger ones may well reach 100 years. Whether this can be substantiated still remains to be seen.

The Nile crocodile ranks as one of the most dangerous of all crocodilians surpassed only by the Saltwater or Estuarine crocodile, *C. porosus*. There are annually many incidents of crocodiles taking people, sometimes even emerging from the water to do so. It is therefore very important to avoid entering waters known to harbour crocodiles. Carelessness is largely responsible for the loss of human life. Crocodiles are not essentially man-eaters in the true sense of the word, but are opportunistic taking whatever comes along. It is difficult to assess how many people are killed by crocodiles as many ascribed deaths may be due to other causes, such as murders and accidents. Most crocodiles are fish eaters and only take to mammals once they have reached 3 to 5 m in length. Of the hundreds of stomachs examined by Cott and his collaborators only four contained human remains; two being from the Kafue River in Zambia, one from the Zambesi and one from Zululand. These could have been deaths resulting from drowning. Tony Pooley in his book *Discoveries of a crocodile man* investigated many attacks by crocodiles on people in northern Zululand and southern Mozambique, most of which, as was to be expected, took place during the warm summer months from November to April. Approximately half of the attacks were fatal, mostly because of the large size of the crocodiles, while the remaining attacks were by smaller individuals, which the victims managed to fight off. Crocodiles may haunt sites of a successful attack in the hope of another, but true man-eaters occur only among the big cats. Despite the potential threat posed by crocodiles, many indigenous people wade through rivers and into lakes while fishing, washing clothes, bathing or simply to get to the other side.

The demise of the crocodile over much of its former distribution in the wild and the concomitant scarcity of crocodilian skins for the leather trade has resulted in the establishment of crocodile and alligator farms and ranches worldwide. South Africa is no exception and close to 40 farms had been established by 1992. However, the uncertain market and fluctuations in the market price of skins has led to the closure of many. In a chapter on 'The status of crocodile farming in South Africa' in the book *'Conservation and utilization of the Nile crocodile in South Africa*, J. Marais and G. Smith reported that only 5 387 skins were exported during 1990, which is a minor portion of the 760 000 skins marketed during the same year worldwide. It seems that the fluctuating skin market is too volatile for the high cost of rearing crocodiles. Despite this, crocodile farming is helping to slow down the illegal harvesting of wild crocodilians.

TOP: *Two- to three-year-old crocodiles ready for slaughter.*
BOTTOM: *Crocodile rearing hothouse.*

What of the future?

In 1978 the first *Red Data Book* on reptiles and amphibians was published and included 34 threatened or endangered reptile species. In 1988 a revision of this was compiled listing 76 threatened or vulnerable reptile species, including one species considered to be extinct. The reasons for this increase in the number of taxa included in the revised book were partly as a result of an improved database but many of the species were incorporated as a result of increased pressure on the environment by a burgeoning human population and associated development. In 10 years the number of affected species has doubled. It is likely that many more taxa will be added in a revision due to their vulnerability, many species having very specific habitat requirements and very local distributions.

Our knowledge of South African reptiles is very fragmented, particularly with regard to their ecological requirements. Much of what has been learnt has been from animals in captivity and incidental observations in the wild. It is very difficult to do field studies on these animals and funding for this and for the specialised equipment required, is extremely limited. Yet without such knowledge of home range size, habitat and food requirements, densities and lifestyle it is almost impossible to guarantee the survival of such animals. We do not know how abundant snakes were when Europeans first arrived here, so that one does not have a baseline against which to measure their current status. The chances of determining the current status of a species is very subjective and based purely on the findings of a few amateur and professional herpetologists who are active in the field and from time to time record its presence. The result is the compilation of a *Red Data Book*, which attempts to highlight those species considered to be vulnerable or threatened as a consequence of their distribution, habitat requirements, abundance and their response to past and current human activities.

Humans' inherent phobia concerning snakes is blown exponentially out of proportion to the threat they pose. In the South Africa of today it is far more likely that one will be murdered, die in a vehicle collision, or from other causes than from a bite of a poisonous snake while walking about in the bush. The adage that the only good snake is a dead snake indicates total ignorance of the role that these animals have in the ecosystem. They are essential to the maintenance of an ecological balance between plants and animals including humans. Snake venom has been in use medicinally for some time to treat various human ailments and it is likely that more will follow.

Many reptile species are catered for in our system of nature reserves, especially those housing the megafauna such as elephant, buffalo and rhino, as these require large areas to sustain viability of populations. This means large protected areas for the reptile species occurring in them. However, such areas have to be justified and managed which results in the construction of infrastructure with their associated impact on reptiles. Firebreaks are burnt or bulldozed and blocks of veld are burnt on a rotational basis. This all impacts on reptiles and we have no means of knowing what effect such management tools have on the lower vertebrates.

However it is mostly those species that occur outside the protection of nature reserves which are of greatest concern. Already it appears that we have lost one lizard species as result of development, while it is likely that species as yet undescribed may be facing the same fate. Local populations of species have been exterminated by agricultural practices. Over large areas of cultivation most reptiles have been decimated with fragmented populations only surviving along road verges. Farmers in several areas have pointed out that common species such as Puff adders have disappeared from their areas attributing this to the use of herbicides and pesticides. This is only part of the picture, what is perhaps more important is that in most areas very little of their habitat remains. The small isolated pockets left do not provide sufficient food or shelter for the continued survival of remnant individuals, which now cannot make contact with other individuals for reproductive purposes and the transfer of genes. Several instances of this have been briefly touched on in the previous chapters and need not be repeated here.

The trend towards urbanisation in South Africa has resulted in additional impacts on the herpetofauna. Mention has been made of some reptile species that can be considered commensal, that is species which are able to adapt and even flourish as a consequence of human activities. A study of the herpetofauna of Durban and the impact of urbanisation has revealed that two species, the Nile crocodile and the Cape file snake have become extinct in Durban while the ranges or numbers of 14 other reptile species were declining. Two other species, the Tropical house gecko and the Flower pot snake, *Rhamphotyphlops braminus*, which did not occur in Durban previously have become established as a result of translocations. Durban is fortunate in having several nature reserves within its municipal boundaries, which have contributed substantially to the existing species richness, but as these become more isolated by increased development, they become islands with a concomitant drop in biodiversity as a result. Areas like Johannesburg and Pretoria where open spaces are on the decline are likely to be much more impoverished.

As traffic increases, roads have become man-made barriers to the movement of animals and contribute extensively towards the decimation of reptile populations. Snakes are routinely seen dead on roads. In 23 surveys of snakes along the same stretch of 108 km of tarmac and 30 km of gravel roads in the Bela Bela district in Limpopo Province over the period December 1994 to October 1997, two amateur herpetologists recorded 170 snakes of 27 species. Of these 125 snakes, or 73,5 per cent, representing 20 species were found dead on the road. The most frequent species killed arranged in order of frequency were Puff adders, Common egg-eaters, Brown house snakes, Mozambique spitting cobras and Herald snakes. These are common species but among the road kills were three Southern African pythons, two Bicoloured quill-snouted snakes and a Cape file snake, all uncommon or rare species. These roads are country roads and not four-lane highways with concomitant traffic volumes where anything crossing the road will be struck before reaching the other side.

Modern man has had a tremendous impact on these animals and continues to do so despite a more in-depth knowledge of our reptiles, not least of which is the capture of animals for the international pet trade. Millions of reptiles and amphibians are traded annually and it is estimated that trade in wildlife and its products is second only to that of the drug trade. This trade mostly targets attractive and appealing species often selecting one sex because of its brilliant colours. Over a nine-year period from 1985 to 1993, more than 250 000 chameleons of various species were traded, most of which went to the USA. Many South African species have been targeted without any studies being done on the sustainability of such harvesting. Harvesting in many instances leads to large-scale habitat destruction, the trade therefore having a dual impact on the species, an untenable situation.

To add to this, the demand for skins, food and medicine, especially in the Far East, is resulting in serious population declines. The scale of this onslaught is unbelievable, with 1,5 million snakes being exported from Thailand to markets in China, Taiwan and Hong Kong for this purpose alone. This in turn resulted in a tremendous increase in the rodent population and a concomitant substantial drop in the amount of rice harvested.

In parts of the Philippines sea snakes gather in large numbers at certain times of the year to reproduce and are opportunistically harvested in very large numbers and exported to China and Japan for food, without any knowledge of the sustainability of such harvesting. It is only a matter of time before many more species are listed as endangered or threatened by exploitation.

South African has been blessed with an unprecedented array of plant and animal species, of which the reptiles form an integral part. This is something to be cherished, not denigrated, ignored or accepted at face value. It is a responsibility that has been handed down, albeit in a much impoverished state. One species is considered extinct with another 16 species considered vulnerable or endangered. It is time to remedy the situation and ensure that no other species follow the same path as that of Eastwood's plated lizard.

Illegal confiscated Leopard tortoises – Quo vadis? Photo: Lorna Stanton.

Bibliography

Alexander, G. 1990. Reptiles and Amphibians of Durban. *Durban Museum Novitates* 15: 1-41.

Archer, W.H. 1967. The Tortoise with a difference. *Afr. Wildl.* 21(1): 59–66.

Archer, W.H. 1967. The Angulated Tortoise. *Afr. Wildl.* 21(2): 137–143.

Archer, W.H. 1967. The Geometric Tortoise. *Afr. Wildl.* 21(4): 321–329.

Bellairs, A. d'A. 1969. *The Life of Reptiles* Vol. 1 & 2. The Weidenfeld and Nicholson Natural History, Weidenfeld and Nicholson, London.

Boycott, Richard C. & Borquin, Ortwin. 2000. *The southern African Tortoise Book: A guide to southern African Tortoises, Terrapins and Turtles.* Published by O. Borquin, KwaZulu-Natal

Branch, W.R. (ed.) 1988. South African Red Data Book – Reptiles and Amphibians. *S. Afr. Nat. Sci. Prog. Rpt* 151: 1–239.

Branch, Bill. 1998. *Field guide to Snakes and other reptiles of southern Africa.* Struik, Cape Town.

Broadley, D.G. 1983. *FitzSimon's Snakes of Southern Africa. Delta Books*, Johannesburg and Cape Town.

Burrage, B.R. 1974. Population structure in *Agama atra* and *Cordylus c. cordylus* in the vicinity of De Kelders, Cape Province. *Ann. S. Afr. Mus.* 66(1): 1–23.

Cott, Hugh B. 1961. Scientific results of an inquiry into the ecology and economic status of the Nile crocodile (*Crocodilus niloticus*) in Uganda and Northern Rhodesia. *Transactions of the Zoological Society of London.* Volume 29 (4): 211–356.

Cowles, R.B. 1930. The life history of *Varanus niloticus* (Linnaeus) as observed in Natal, South Africa. J. Ent. & Zool. 22: 3–21.

Gans, C. 1969. Amphisbaenians – reptiles specialized for a burrowing existence. *Endeavour* 28: 146–151.

Haacke, W.D. 1969. The call of the Barking Geckos (Gekkonidae: Reptilia). *Scient. Pap. Namib Desert Res. Stn* 46: 83–93.

Haacke, W.D. 1975. The Burrowing Geckos of southern Africa 1 (Reptilia: Gekkonidae). *Ann. Tvl Mus.* 29(12): 197–243.

Hughes, G. 1982. *Sea Turtles: a guide*. Natal Parks Board, Pietermaritzburg.

Isemonger, R.M. 1955. *Snakes and snake-catching in Southern Africa.* Howard Timmins, Cape Town.

Jacobsen, N.H.G. 1982. The ecology of the reptiles and amphibians of the *Burkea africana* – *Eragrostis pallens* savanna of the Nylsvley nature reserve. Unpubl. MSc thesis, University of Pretoria, Pretoria.

Jacobsen, N.H.G. 1985. *Ons Reptiele*. CUM Boeke, Roodepoort.

Jacques, Jeanne. 1966. Some observations on the Cape terrapin. *Afr. Wildl.* 20(2): 137–150.

Livingstone, David. 1858. *Missionary travels and researches in South Africa.* Harper & Brothers, New York.

McLachlan,G.R. 1978. South African Red Data Book – Reptiles and Amphibians. *S. Afr. Nat. Sci. Prog. Rpt* 23: 1–53.

Pienaar, U. de V., Haacke, W.D. & Jacobsen, N.H.G. 1983. *The reptiles of the Kruger National Park*. National Parks Board of Trustees of the Republic of South Africa.

Pitman, Capt. C.R.S. 1974. *A guide to the snakes of Uganda*. Wheldon and Wesley, Codicote.

Pooley, Tony. 1982. *Discoveries of a Crocodile man*. William Collins, Johannesburg, London.

Pritchard, Peter H. 1979. *Encyclopedia of Turtles*. T.F.H. Publications.

Ross, Charles A. (ed.) 1989. *Crocodiles and Alligators*. Merehurst Press, London.

Smith, G.A. & Marais, J. (eds) 1992. *Conservation and utilization of the Nile crocodile in South Africa: Handbook on Crocodile farming*. 186pp.

Sweeney, R.C.H. 1961. *Snakes of Nyasaland*. The Nyasaland Society and the Nyasaland Government. Govt. Printers, Zomba, Nyasaland.

Vogel, Zdenek. 1964. *Reptiles and Amphibians: their care and behaviour*. Studio Vista, London.

Glossary

adrenal gland	a gland producing adrenalin which is vital in initiating escape and defence responses
Aglypha	snakes without poison fangs
acrodont	arrangement of teeth along the top ridge of the jawbone
allophores	red pigment cells
amniotic	refers to the protective membrane, the amnion, surrounding the embryo
Anapsida	a grouping of reptiles in which the skull does not exhibit temporal openings or foramen
Anthracosaurian	refers to the line of primitive amphibians regarded as the ancestors of the reptiles
annulus	rings of scales encircling the body of worm lizards (plural: annuli)
aposematic	a term referring to highly contrasting colours such as black, red, white or yellow found on some animals warning potential predators that they may be poisonous or distasteful
arachnids	general term for spiders, mites, ticks and scorpions, etc.
Archosauria	animals such as crocodiles in which the skull exhibits two temporal openings
binocular	refers to the field of vision of both eyes in front of the animal, enabling it to even see immobile prey
brille	refers to a transparent scale formed by both upper and lower eyelids growing together and covering the eyes in some reptiles
caudal	refers to the tail
cervical	refers to the neck
chromosome	sites of inherited material found within cells
cladoic	refers to an egg which is enclosed in a shell; only an interchange of gases takes place through the walls
cloaca	a general vestibule within which the openings of the reproductive organs, urinary tract and intestines are found, which opens to the outside via the anus
condyle	a rounded knob on which the skull and bones articulate
cryptic	hidden, obscure or camouflaged
cytotoxin	a toxin which destroys tissue cells
dermal	the sub-epidermal layer
dimorphism	different appearance of male and female in colour or morphology

ectothermic	refers to the reliance of reptiles to an outside source of heat
endemic	only occurring locally, i.e. in a country, province, farm etc.; not found naturally elsewhere
era	refers to a specific pre-historical period of time
Eustachian tube	forms part of the inner ear in animals and humans
exotic	alien
femoral pores	glandular pores on the underside of the thigh found mostly in males in some lizard species
fossil	preserved solidified remnants of animals
fossorial	refers to an animal having a burrowing existence
fovea centralis	a small area on the retina formed by slender cone cells which allow acute sight
guanophores	cells containing colourless crystals
haemotoxic	refers to poisons toxic to blood cells
hemipenes	paired reproductive organs of male lizards and snakes
hierarchy	refers to a social ranking order with a dominant individual, usually a male at the head
interclavicle	a dermal bone found anterior to the sternum or breastbone, a primitive characteristic
keratin	a tough waterproof protein covering produced by the epidermis
lamellae	scale rows under the toes and some tails of some gecko species
lamina	shields of keratinised material covering the shells of tortoises, terrapins and turtles
Lepidosauria	a grouping of primitive scaly reptiles
lepidosis	refers to the arrangement of scales on the body of lizards and snakes
lipophores	yellow pigment cells
maxilla	bone in the upper jaw
melanistic	blackish in appearance
melanophores	black pigment cells
monotypic	one of a kind
musk glands	glands producing a substance (musk) with a strong scent
neurotoxic	toxins affecting the nervous system; a nerve poison affecting nerve impulse transmission
Opisthoglyphs	refers to a group of snakes in which the poison fangs are situated below or diagonally below the eye

ossification	bone forming
osteoderms	bone-like supports occurring in the scales of some lizards and all crocodiles
oviparous	egg-laying
ovoviviparous	a term referring to the almost full term retention of the eggs in the oviducts before they are laid throughout incubation, the eggs being laid with a more or less fully developed embryo, which are laid a few days before hatching
parallel evolution	the appearance and habits of unrelated animals that have evolved similar to one another but on different continents
parthenogenesis	refers to reproduction by a female parent only, without fertilisation
pineal eye	a primitive character which is still present in some lizards, especially iguanids; it appears to assist in thermoregulation
pituitary gland	a gland producing many hormones, situated between the roof of the mouth and the floor of the brain
pleurodont	refers to the teeth mounted along the inside of the jawbone
polyphyletic	origins are derived from more than one ancestral line
premaxilla	a bone at the front of the upper jaw
protozoan	single cellular animal
pterygoid	a bone in the upper jaw
quadrate bone	a bone at the corner of the upper jaw
retinal cones	important component of the eye involving sight
rostral	a scale at the tip of the snout
rupicolous	rock living
synaps	a gap between nerve endings which a nerve impulse must bridge
temporal	refers to the region on the side of the head between the eye and ear or is time related depending in which context it is referred to
territory	an area in which an animal lives and which it defends against others of the same species
terrestrial	ground living
ultraviolet rays	an invisible part of sunlight characterised by long wave lengths
vestigial	describes an organ that has degenerated or atrophied, having become functionless in the course of evolution
viviparous	live bearing, characterised by a placental bond between mother and embryo

INDEX

A

Acanthocercus atricollis 41
Acontias 58
    plumbeus 58
Acontophiops 58
    lineatus 58
adder, Berg 123
    Common night 81, 127
    Demon 127
    European 122
    Gaboon 126, 127
    Horned 5, 123
    Peringuey's 122, 123
    Puff 10, 80, 86, 87, 108, 124, 125, 126, 141
    Schneider's 123
    Snouted night 127
Aelurognatus 1, 2
Aethomys chrysophilus 127
Afroedura 53
    multiporis 53
Agama agama aculeata 45
    agama distanti 45
    agama atra 46
    agama knobeli 46
agama, Ground 45
    Distant's ground 45, 46
    Rock 46
    Tree 41, 44
Agamidae 44
Aglypha 13, 14
alligators 10
Amblyodipsas 93
Amblyodipsas polylepis 93
Amblyomma 75
Amphisbaenia 75, 76
Amphisbaenidae 75
Amplorhinus 95
    multimaculatus 107
anaconda 86
Anapsida 3
Androstachys johnsoni 77
Angolosaurus skoogi 71
Anochaetus faurei 69

ants, Driver 88
    Matabele 88
Aparallactus 93
    capensis 94
Archosauria 3
asp, Bibron's burrowing 94
    Duerden's 94
Aspidelaps lubricus 118
    scutatus 118
Atractaspididae 92, 93
Atractaspis 93
    bibronii 94
    duerdeni 94
Atilax paludinosus 135
Aulacephalodon 1, 2
Australopithecus 2

B

barking gecko, Carp's 47
    Common 47, 48
    Koch's 47
beetle, Darkling 72
    Tiger 63
    Toktokkie 72
Bitis 122
    arietans 80, 124
    atropos 123
    caudalis 123
    gabonica 126
    peringueyi 122
    schneideri 123
Bloukopkoggelmander 44
boa constrictor 10
Boidae 12, 89
Boomslang 56, 97, 98, 104, 105, 106, 118
Bradypodion 54
    transvaalense 56
bullfrog, Giant 86
burrowing skink, Limpopo 57
    Montane 57
Bushmaster 11
Bushpig 135

C

Carabidae 63
Caretta caretta 36
Causus defillippi 127
    rhombeatus 127
centipede-eater, Cape 94
Chamaeleo 54
    dilepis 7, 42
    namaquensis 57
Chamaeleonidae 54
chameleon, Common flap-necked 7, 10, 12, 42, 43, 54, 55, 56
    Namaqua 56, 57
Chelonia 2, 8, 20
Cheloniidae 36
Chelonia mydas 36
Chersina angulata 29
Chirindia langi 76, 77
Chiromantis xerampelina 98
Chondrodactylus angulifer 52
cobra, Black spitting 114
    Black-necked spitting 113, 114
    Cape 6, 86, 108, 111, 112
    Egyptian 111
    Forest 112
    King 111
    Mozambique spitting 113, 114, 141
    Snouted 86, 111
Colubridae 95
Cordylidae 5, 66
Cordylosaurus subtessellatus 71
Cordylus aenea 68
    catapharctus 66
    giganteus 66
    jonesi 67
    macrolepis 68
    macropholis 67
    melanotus 68
    tasmani 67
    vittifer 67
    warreni depressus 67
    warreni 67
Cottonmouth 121
Cotylosaur 2
crag lizard, Drakensberg 68
crocodile, Nile 11, 128, 130, 131, 136, 139
    Saltwater 136
    Siamese 136
Crocodilia 2, 128
Crocodylidae 129
Crocodylus niloticus 9, 128, 130
    porosus 136
    siamensis 136
Crocuta crocuta 135
Crotaphopeltis hotamboeia 79, 87, 103, 104
Crotalus cerastes 122
Cryptactites peringueyi 53
Cryptoblepharus boutoni 18, 61
Cryptodira 17, 23, 25, 33
Cycloderma frenatum 32

D

Dalophia pistillum 77
Dasypeltis inornatus 102
    medici 102
    scabra 79, 102
day gecko, Namaqua 53, 54
Deinosuchus 130
Dendroaspis angusticeps 98, 116
    polylepis 83, 87, 116
Dermochelys coriacea 34
desert lizard, Shovel snouted 64
    Smith's 64
    Wedge snouted 64, 65
Dispholidus typus 98, 104
Dorylus sp. 88
dragon, Komodo 71
Duberria lutrix 96
    variegata 97
dwarf chameleon, Transvaal 38, 56
dwarf gecko, Black-spotted 49
    Bradfield's 49
    Common/ Cape 38, 41, 48, 49, 50
    Granite 49
    Methuen's / Woodbush 50
    Ocellated 49, 50
    Stevenson's 49
    Waterberg 49, 50

E

eagle, Martial 75
Echis carinatus 122
egg-eater, Brown 102
    Common 79, 102, 110
    East African 102
Elapidae 17, 110
Elapsoidea boulengeri 120
    sundevallii 120
Eretmochelys imbricata 21
Eumeces 61

F

Fer de lance 111
flat lizards 66, 68
flat gecko, African 10
    Woodbush 53
fly, Tsetse 85
frog, Grey tree 98

G

Galarella sanguinea 85
gecko, Wahlberg's velvety 47, 51
    web-footed 10, 11
        Kaoko web-footed 47
        Namib 47
Gekkonidae 47
Geochelone 20
Geochelone gigantea 27
    pardalis 25
Gerrhosauridae 5, 69
Gerrhosaurus flavigularis 43, 70
    major 43, 70
    validus 6, 43, 69
Gigantophis 89
girdled lizard, Armadillo 66
    Dark 67
    Giant 66
    Jones' 67
    Large-scaled 67
    Tasman's 67
    Transvaal 67
    Warren's 67
Gondwanaland 17, 54
goose, Egyptian 91
Grammomys dolichurus 127
grass lizard 66
    Large scaled 68
    Transvaal 68
Grewia 70
ground gecko, Giant 10, 41, 52
guineafowl 91

H

Hamerkop 117
Heliobolus lugubris 63
Hemachatus haemachatus 86, 113
Hemidactylus mabouia 38, 50
hinged terrapin, Mashona 25
    Pan 25
    Serrated 24, 25
    Yellow-bellied 25
hinged tortoise, Bell's 31

    Lobatse 31
    Natal 31
    Speke's 31
Holaspis guentheri 62
Homo erectus 2
Homo sapiens 2
Homopholis mulleri 52
    wahlbergi 51
Homopus 20
Homopus areolatus 30
    boulengeri 30
    femoralis 30
    signatus 30
Homoroselaps dorsalis 119
    lacteus 119
house gecko, Tropical 10, 38, 50, 139
Hyaena brunnea 75
hyena, Brown 75
    Spotted 135

I

Ichnotropis capensis 63, 64, 65
    squamulosa 64

J

Jararacussu 111

K

Kaokogecko vanzijli 47
Kinixys belliana 31
    lobatsiana 31
    natalensis 31
    spekii 31

I

land snail, Giant 51
Lacerta lepida 62
    vivipara 64
Lacertidae 62
Lacertilia 38
Lamprophis aurora 100
    fiskii 99
    fuliginosus 79, 99
    fuscus 99
    guttatus 100
    swazicus 99
Lanthonotidae 78
leaf-toed gecko, Peringuey's coastal 53
Lebombo ironwood 77
legless skink, Gariep 59
    Striped blind 57, 58

Giant 57
Woodbush 58
Lepidochelys olivacea 37
Lepidosauria 3
Leptoptilus crumeniferus 135
Leptotyphlopidae 12, 89
Leptotyphlops longicauda 89
Liasis amethystinus 90
lizard, Bushveld 63
    Fringe-tailed 62
    Jewelled 62
    Viviparous 64
Lycodonomorphus laevissimus 98
    obscuriventris 98
    rufulus 98
Lycophidion capense 101
    variegatum 101
Lycosid spider 41
Lygodactylus 47
    bradfieldi 49
    capensis 38, 48
    methueni 49
    nigropunctatus 49
    ocellatus 49
    ocellatus soutpansbergensis 50
    stevensoni 49
    waterbergensis 49
Lygosoma sundevalli 59

M

Macrotermes 117
mamba, Black 8, 16, 83, 85, 86, 87, 111, 116, 117, 118
    Green 97, 98, 116, 117, 118
Marula 52
Mehelya capensis 96
Meroles anchietae 64, 65
    ctenodactylus 64
    cuneirostris 64
Mesosaurus 2
mice, Multimammate 125
    Striped field 125
    Woodland 127
moccasin, Water 121
mongoose, Banded 85, 86, 135
    Slender 85, 86
    Water 137
monitor, Veld 9, 19, 43, 71, 72, 74, 75
    Water 9, 13, 43, 71, 72, 73, 74
Monopeltis 76
    infuscatus 76, 77
Montaspis 95

Mopane 77
mountain lizard, Cottrell's 39
Msimbiti 77
Mungos mungo 135

N

Naja a. annulifera 111
    haje 111
    melanoleuca 112
    mossambica 113
    naja 113
    nigricollis nigricincta 114
    nigricollis woodi 113
    nivea 112
Nucras holubi 62
    intertexta 62
    lalandei 62

O

Opisthoglypha 13, 14
Otomys spp. 25, 127

P

Pachydactylus 50
    affinis 51
    bibroni 51
    maculata 51
    turneri 51
padloper, Greater 30
    Karoo 30
    Lesser 30
    Southern speckled 30
Palmatogecko rangei 10, 47
Panaspis 60
    maculicollis 60
    wahlbergi 60
Pangaea 17
Pelamis platurus 121
Pelomedusidae 24
Pelomedusa subrufa 24
Pelusios castaneus 25
    rhodesianus 25
    sinuatus 25
    subniger 25
Phacocoerus aethiopicus 135
Phelsuma ocellata 54
Philothamnus hoplogaster 97
    natalensis 97
    semivariegatus 85, 97
plated lizard, Dwarf 71
    Eastwood's 69, 70

Five toed 69, 70
Giant 1, 13, 69
Namib 71
Rough-scaled 70
Yellow-throated 43, 70, 71
Platysaurus 68
guttatus 43
Pleurodira 17, 23, 24
Polemaetus bellicosus 75
Potamochoerus porcus 135
Praomys natalensis 125
Prolacerta 1, 2
Prosymna 95, 96
jani 96
sundevalli lineata 95
sundevalli sundevalli 95, 96
Proteroglypha 13, 14
Psammobates 20
geometricus 20, 27
oculiferus 29
tentorius 28
trimeni 29
Psammophis 109
brevirostris 110
crucifer 109
jallae 109
subtaeniatus 109
Psammophylax rhombeatus 108
tritaeniatus 83, 109
Pseudaspis cana 101
Pseudocordylus 68
Ptenopus carpi 47
garrulus 47
kochi 47
Python anchietae 90
natalensis 9, 82, 90
regius 90
reticulatus 90
sebae 90
python, African 90
Amethystine 90
Ball 90
Reticulated 86, 90
Southern African 9, 13, 82, 90, 91, 92
Pyxicephalus adspersus 86

Q

quill-snouted snakes 93

R

rat, Red veld 127
Vlei 125
Rhabdomys pumilio 125
Rhamphotyphlops braminus 16, 89, 139
Rhinotyphlops schlegeli 88
Rhynchocephalia 2
Rinkhals 66, 86, 113, 114, 115, 121
rough-scaled lizard, Cape 63, 64, 65

S

sandveld lizard, Delalande's 62
Holub's 62
Spotted 62
Sarchosuchus 130
Sauria 38
Scelotes 57
limpopoensis 57
mirus 57
Scincidae 57
Sclerocarya birrea 52
Scopus umbretta 117
Shovel-snout, Mozambique 96
Sundevall's 95, 96
sidewinder, American 122
skaapsteker, Spotted 84, 108
Striped 71, 83, 108
skink, Black-lined 59, 60
Blue-tailed koppie 60
Cape 61
Eastern coastal 60
Striped 60
Sundevall's writhing 59
Variable 17, 61
slug-eater, Common 96, 107
Variegated 97
snake, American coral 6, 111
Aurora house 100
Beetz's tiger 104
Bibron's blind 88
Bicoloured quill-snouted 93, 139
Black file 96
Black water 98
Boulenger's garter 120
Brown house 79, 99, 110, 141
Brown water 98
Cape file 86, 96, 139

Cape wolf 100, 101
Common purple-glossed 93
Coral 118, 119
Cross-marked sand 109
Dwarf crowned 111
Dwarf crowned sand 109
Eastern tiger 104
Fierce 111
Fisk's house 99
Flowerpot 16, 89, 139
Green water 78, 97
Herald 79, 86, 87, 103, 104, 110, 139
Jalla's sand 109
Long-tailed thread 89
Mole 100, 101, 102, 107
Natal green 97
Red-lipped 79, 110
Reed 107, 108
Schlegel's blind 88
Shield-nosed 5, 118, 119
Short-snouted sand 110
Sonoran king 6
Spotted bush 85, 97, 98
Spotted harlequin 119
Spotted house 82, 100, 101
Stripe-bellied sand 109
Striped harlequin 119, 120
Sundevall's garter 120
Swazi house 99
Transvaal quill-snouted 93
Variegated wolf 100, 101
Vine 83, 85, 87, 104, 105, 106, 107
Western black-striped 111
White-lipped water 98
Yellow-bellied house 99
Yellow-bellied sea 121
Zebra 114
snake-eyed skink, Bouton's 18, 61
　　Spotted-neck 60
　　Wahlberg's 60
Squamata 2, 16, 38
Solenoglypha 14
stork, Marabou 135
Suncus varilla 95
Sungazer 66

T

Taipan 111
Telescopus beetzii 104
　　semiannulatus semiannulatus 104
　　semiannulatus polystictus 104

termite, Harvester 53
terrapin, American red-eared 18, 32
　　Cape 6, 24
　　Nile soft-shelled 32
　　Zambesi soft-shelled 32
Testudinidae 25
Tetradactylus africanus 69
　　eastwoodae 69, 70
Thelotornis capensis 83, 87, 105
thick-toed gecko, Bibron's 51
　　Spotted 51
　　Tiger 51
　　Transvaal 51
　　Turner's 10, 12, 51
　　Van Son's 51
Thrinaxodon liorhinus 1, 2
tortoise, Angulate 29
　　Geometric 20, 27, 28
　　Giant 27
　　Kalahari serrated 29
　　Knoppiesdop 28
　　Leopard 17, 18, 25, 26, 27
　　Mountain 25
　　Parrot-beak 30
　　Ploughshare 29
　　Tent 28
Trachylepis (= *Mabuya*) *capensis* 61
　　depressa 60
　　margaritifer 60
　　punctatissimus 60
　　striata 60
　　varia 40, 61
Tracheloptychus madagascariensis 71
Trachemys scripta 32
Trionyx triunguis 32
Tropidosaura cottrelli 39
Tuatara 2, 11, 17
turtle, Green 33, 34
　　Hawksbill 21, 33, 34
　　Leatherback 33, 34, 36
　　Loggerhead 33, 34, 36, 37
　　Olive Ridley 33, 34, 37
Typhlopidae 12, 88
Typhlops bibronii 88
Typhlosaurus 58
　　aurantiacus 59
　　gariepensis 59
　　lineatus 58
　　richardi 59
　　subtaeniatus 59

V

Varanidae 71, 78
Varanus albigularis 9, 43, 71,
    komodoensis 71
    niloticus 9, 43, 71
viper, Russell's 122
    Saw-scaled 122
Vipera 122
    berus 122
    russelli 122
Viperidae 110, 121

W

Warthog 137
worm-lizard, Blunt-tailed 77
    Dusky spade-snouted 76, 77
    Kalahari round-headed 76
    Msimbiti 76, 77

X

Xenocalamus 93
    bicolor 93
    bicolor australis 93
    bicolor bicolor 93
    bicolor lineatus 93
    transvaalensis 93

Z

Zygaspis 77
    quadrifrons 77